Lab Manual
to accompany

Chemistry in Your Life

Second Edition

Brian T. Arneson
University of Texas at Austin

Anna R. Bergstrom
Texas Lutheran University

W. H. Freeman and Company
New York

ISBN: 0-7167-6956-5
EAN: 97807167-6956-9

Printed in the United States of America

First printing

W. H. Freeman and Company
41 Madison Avenue
New York, NY 10010
Houndmills, Basingstoke RG21 6XS England
www.whfreeman.com

Table of Contents

Paper Chromatography of Inks and Dyes

Introduction

Many common materials around us are mixtures of various types of elements and compounds. The properties of a mixture are usually a combination of the properties of its individual parts. Mixtures can be separated into their component parts. The process of separating the components of a mixture can be as simple as using a filter or mesh screen to separate larger solids from a mixture. However, not all mixtures are separated so easily. Solutions are mixtures in which one species is dissolved in a liquid, called the solvent. Salt water is a solution prepared by dissolving salt in water. Separating the components in salt water requires boiling, or evaporating, the water component of the mixture, leaving the salt behind.

Colored dyes and inks used in markers are another example of a solution. The specific color of the dye comes from large molecules containing 20, 30, or more atoms. The molecules are dissolved into a solvent to create the solution that is put into the marker. When the marker is pressed against a sheet of paper, the solution is drawn through the tip of the marker and onto the paper. The solvent evaporates, leaving the colored molecules behind for us to see. There are many suitable molecules that can be used as dyes. Also, using more than one type of dye molecule in a solution can produce different colors.

This experiment uses paper chromatography to investigate the different dyes contained in felt-tip markers.

Background Information

Separating the individual components of a solution takes some knowledge of the physical properties of the components. A common method of separating the components in a solution is chromatography. Chromatography takes advantage of different properties, like solubility and polarity, possessed by the molecules that make up the mixture. These properties are related to the shapes, structures, and bonding of the molecules. In this experiment you will use paper chromatography to separate the dyes used in felt-tip markers.

The paper used for paper chromatography is a very pure and absorbent variety of paper. The mixture that is to be separated is applied to the paper as small spots and then the paper is dipped into a carefully designed solvent that may be a single liquid or a mixture of liquids. The solvent rises up the paper by capillary action and is called the mobile phase. As the solvent passes over the spot of the substance to be separated, the components in the mixture that are soluble in the solvent move up the paper with the solvent. In order to be an effective separation, the components of the mixtures will have a

varying degree of solubility in the solvent. Components that are most soluble in the mobile phase are carried up the paper fastest and will end up at the top of the paper. Components that are least soluble in the mobile phase will remain near the bottom of the chromatography paper.

The solubility of a compound in a solvent is related to the polarity of the compound and the polarity of the solvent. Compounds can be classified as polar, non-polar, or a couple of other classifications like metallic or ionic. The details of the classification is beyond the scope of this laboratory, but it is useful to know that these labels describe the charge distribution and bonding within a compound. The degree of charge separation in a compound affects the degree of polarity. Compounds can vary from extremely polar to almost non-polar. Solvents, likewise, can be polar or non-polar. The solubility of a compound in a solvent generally obeys the rule "like dissolves like." In other words, polar compounds will dissolve in polar solvents, and non-polar compounds will dissolve in non-polar solvents. Within the context of chromatography, using solvents with different polarities affects the separation. When using a polar solvent like water, polar compounds will travel up the paper faster than less polar or non-polar compounds. In contrast, when a non-polar (or less polar) solvent is used, polar molecules will not travel up the paper very much at all.

To compare compounds in a paper chromatography separation, the ratio of the distance a compound moves relative to the distance the solvent travels up the paper is calculated. This ratio is called an R_f value. Figure 1 shows an example of a paper chromatography separation. The distance a compound moves is measured from the baseline, where the original sample is applied, to the midpoint of the spot that has traveled up the chromatography paper. The distance traveled by the solvent is measured from the baseline to the point where the solvent stopped moving up the paper. It is important for this calculation that the paper is removed from the solvent before the edge of the solvent reaches the top of the paper. The distance measurements can be made in any units, so long as the units are the same in each distance measurement!

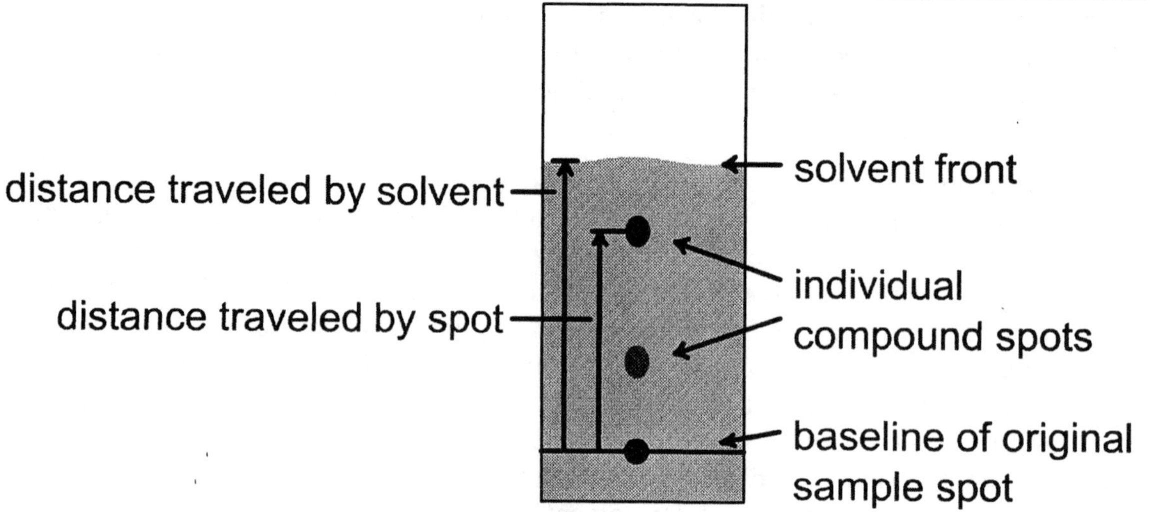

Figure 1. Paper chromatography example.

The R_f value for the compound that traveled the furthest is calculated in the following manner:

Distance traveled by compound: 2.9-cm
Distance traveled by solvent: 3.80-cm
R_f value: 2.9-cm/3.8-cm = 0.76

By careful observation and measurement, you should be able to determine whether the dyes used in markers are pure substances or mixtures of different dyes. Also, you should be able to determine whether markers from different manufacturers use the same dyes.

Chemicals and Equipment

- chromatography paper or high quality filter paper
- 2 test tubes—large enough to hold the chromatography paper assembly
- test tube holder
- 2 corks or rubber stoppers
- 2 small paper clips
- hole punch
- pencil
- ruler
- 2 marker pens—1 water soluble and 1 permanent
- water
- rubbing alcohol

Safety Concerns

- **Use general laboratory safety rules.**

Procedure

1. Clean two test tubes. Label one A and fill it to a depth of approximately 1-cm with water (the **first** solvent). Label the second test tube B and fill it to a depth of approximately 1-cm with rubbing alcohol (the second solvent). The depth of 1-cm will assure that the colored spots you apply later to the chromatography paper will be above the level of the solvent. Set the test tubes aside in a test tube holder while you prepare the paper chromatography strips.
2. Obtain two strips of chromatography paper. Handle the strips by the top end only and handle them as little as possible. Measure about 1.5-cm from the bottom end, which you have not touched, and draw a **light pencil line** on each strip. This line is called the baseline and is the starting point for your R_f measurements.
3. Punch a single hole at the top of the strip. It is from this hole that you will hang your strip in the test tube. Make a hanger from a cork and small paper clip as illustrated in the drawing below. The strip, cork, and paper clip will need to fit in

the test tube and hang freely—measure carefully to insure the baseline is above the solvent level.

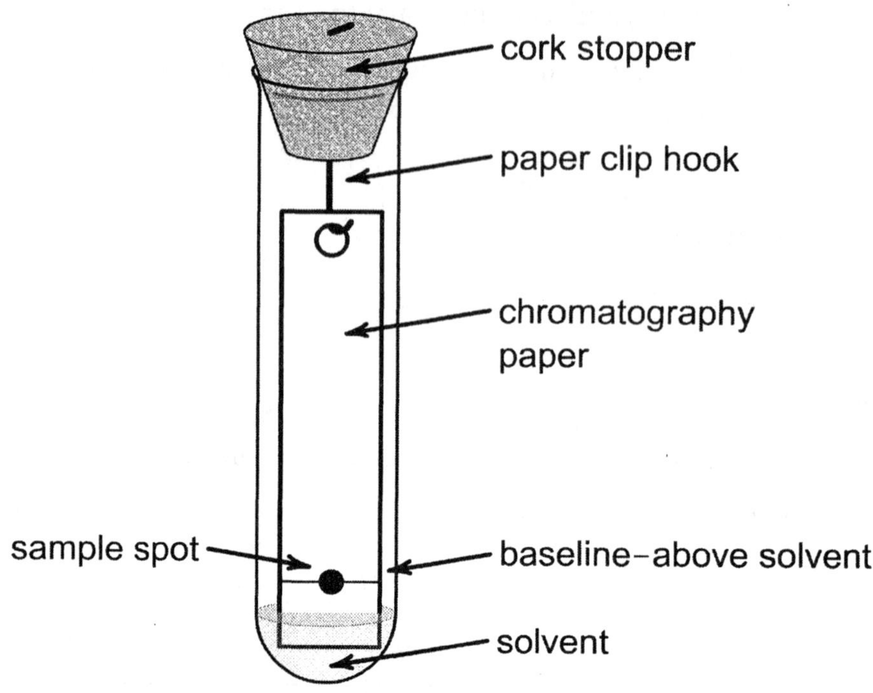

Figure 2. Paper chromatography equipment assembly.

4. Obtain two felt-tip pens. Make sure that one of the pens is water soluble and the other is a permanent pen, which will be alcohol soluble. Apply a drop of one of the ink types at the baseline of each strip. Using a pencil, record on the strip the identity of the pen (water or alcohol soluble). Be careful to keep the spots smaller than approximately 2-mm in diameter. Allow the ink to dry and then reapply more ink. Carefully place the water soluble sample strip into the test tube labeled A and the alcohol-soluble sample strip into the test tube labeled B.
5. Place the test tubes in a test tube holder. It is essential that the test tubes be upright and completely vertical. Watch the developing chromatograms and remove the strip when the solvent is approximately 1-cm from the hole punched in top of the chromatography paper. It may take up to 20 minutes for the solvent to travel up the chromatography paper. You must attentively watch your experiment.
6. When the solvent has reached the top, carefully remove the strips and allow them to air-dry.
7. When the chromatography paper has completely dried, measure and record in a data table the distances needed to calculate the R_f values as described in the background information. If you have more than one developed spot, measure each

spot and calculate R_f values for each. Next to your data table, sketch the chromatography paper with the different spots that have developed.

Disposal and cleanup

☞ All disposable materials can be flushed down the sink or thrown away in the trash.

Data and Observations

General observations

For each chromatography paper, measure the distances from the baseline to the spots representing individual compounds. Record these measurements in data tables like the one shown below. Make a drawing of the chromatography paper next to the data table.

Data table

Marker description			
	Spot 1	Spot 2 (if necessary)	Spot 3 (if necessary)
Distance from baseline to spot			
Distance from baseline to edge of solvent			
R_f value			

Marker description			
	Spot 1	Spot 2 (if necessary)	Spot 3 (if necessary)
Distance from baseline to spot			
Distance from baseline to edge of solvent			
R_f value			

Post Experiment Questions

1. Why was it important to make sure the level of the solvent was below the spot on the chromatography paper?
2. What can you say about the properties of mixtures of colored dyes?
3. Compare your findings to those of your classmates. Are the dyes used in different markers the same?
4. The property tested in this experiment was solubility. Prepare a list of the different dyes used in the markers in your class. Order the list in terms of solubility in the solvent.
5. What would have happened if you had placed the water soluble-marker-spotted and permanent marker-spotted-chromatography papers in the wrong test tubes? If time permits, carry out this experiment to find out if the results confirm your hypothesis.

Information for Instructors

Have a variety of water-soluble and permanent markers on hand, black markers from multiple manufactures or different colors from the same manufacturer. You can extend the focus of this lab by having students propose strategies to test the different dyes used in the markers of the same brand, or in different brands of markers.
The water-soluble markers tend to give a more "blurred" spot than the permanent markers. One step to aid in reducing the blurriness of the developed spots is to allow the initial marker spot to dry before dipping it into the solvent in the test tube.

MYSTERY POWDERS

Introduction

The objective in this experiment is to determine the identity of nine white powders that have been labeled A through I. These nine samples are common household chemicals.

Compound	Formula	Common Source
sodium chloride	$NaCl$	uniodized table salt
sodium bicarbonate	$NaHCO_3$	baking soda
sodium carbonate	Na_2CO_3	washing soda
sodium hydroxide	$NaOH$	lye or drain opener
boric acid	H_3BO_3	roach bait
calcium sulfate	$CaSO_4$	plaster of paris
calcium carbonate	$CaCO_3$	white chalk
cornstarch		cornstarch
sucrose	$C_{12}H_{24}O_{12}$	table sugar

Cornstarch is a mixture of two different compounds and therefore does not have an exact chemical formula.

To accomplish this objective you have access to four reagents, as well as water. The reagents are also common household chemicals.

Reagent	Common Source
litmus	litmus paper
50% isopropyl alcohol	50% rubbing alcohol
iodine solution	tincture of iodine
acetic acid	white vinegar
water	tap or distilled

Background Information

Chemical and physical properties make every substance unique. Physical properties such as color, density, and shape are easily seen with our senses. However, determining some physical properties, like solubility, or chemical properties requires other solutions or chemicals to be used. Reagents are chemicals—either in solid form or dissolved in solution—that are used for specific purposes. In this experiment, there are five reagents; each reagent will be used to help identify one of the nine mystery powders based on solubility, reactivity, or acid/base nature.

Solubility is defined as the amount of a substance that dissolved in a given quantity of a solvent at specific conditions such as temperature and pressure. Of the nine

mystery powders, three will be insoluble in water and six will be water soluble. At the end of the experiment, you will be left with two mystery powders that were originally water soluble. At this point you will vary the temperature of the water. One will be more soluble in hot water and one less soluble in hot water. It is a subtle difference. You will also test for solubility in alcohol during the procedure.

Vinegar, a source of acetic acid, reacts with carbonate to form a gas, carbon dioxide. When vinegar reacts with a carbonate-containing compound, bubbles will be generated. To test a substance for the presence of carbonate, add a couple of drops of vinegar to a small amount of the substance.

Iodine reacts with starch to produce a blue colored solution. If starch is present, the normally brown iodine tincture will turn blue. If starch is not present, the brown color will remain. To test for the presence of starch, drops of iodine tincture are added directly to a small sample of the substance.

Litmus paper is used to determine if a substance is acidic or basic. Litmus paper is available in two colors, red and blue. Red litmus will remain red in the presence of an acid, but it will turn blue in the presence of a base. To test the acid/base nature of a substance using litmus paper, a small amount of the substance is dissolved in water and a drop of the resulting solution is placed a piece of the litmus paper.

This experiment has been designed so that you will report your data and observations in a flow chart. A partially completed flow chart is provided in the data and observations section. The flow chart starts at the top of the page with labels—A through I—for all nine powders. Going down the chart, two empty boxes are drawn beneath each test you will perform. Fill in the boxes with the appropriate labels based upon the results for each test. At the end of the experiment, you should be able to identify the mystery powders according the to tests you perform. Prior to conducting the experiment, you should research a few properties of each common household substance. For example, NaOH is soluble in water, it will not react with iodine, it will not react with vinegar to produce bubbles and it is a base. By researching the properties prior to conducting the experiment you can easily identify the samples at the end of the experiment.

Find the following properties for each of the nine mystery powders. Information is readily available on the Internet and in chemical reference books.

Substance	Water solubility	Alcohol solubility	Reaction with vinegar	Reaction with iodine	Reaction with litmus paper
NaCl					
$NaHCO_3$					
$CaSO_4$					
$CaCO_3$					
NaOH					
H_3BO_3					
Na_2CO_3					
sucrose					
cornstarch					

Chemicals and Equipment

- 17 small test tubes
- 1 test tube rack
- 1 spatula or small scoop
- 4 droppers or beryl-style pipettes
- small sample of nine mystery powders labeled A through I
- red litmus paper
- blue litmus paper
- water, tap or distilled
- vinegar
- iodine
- 50% rubbing alcohol
- tincture of iodine

Safety Concerns

- **Sodium hydroxide is a strong base and is caustic. Since you do not know which unknown is sodium hydroxide, treat all unknown mystery powders with extreme care. If by accident you get any unknown mystery powder on your skin, wash with copious amounts of water and notify your instructor immediately.**
- **Boric acid is mildly dangerous. Since you do not know which unknown is boric acid, treat all unknown mystery powders with extreme care. If by accident you get any unknown mystery powder on your skin, wash with copious amounts of water and notify your instructor immediately.**
- **Goggles should be worn at all times.**

Procedure

1. Obtain nine small test tubes and label them A through I.
2. Add a pea-sized amount of each of the mystery powders to the labeled test tubes.
3. The first test you will perform is to determine if the mystery powders are soluble in water. Using a dropper or beryl-style pipette, add two milliliters of water to each test tube. Gently shake or flick the test tube. Record your observations in the flow chart by writing the labels in the box under either the word soluble or the word insoluble. As a hint to get you started, you should have three mystery powders that are insoluble and six that are soluble.
4. Continuing with the three mystery powders that are insoluble, now you will test for the presence of starch using iodine tincture. You need to add a pea-sized amount of each of these unknowns to three clean and labeled test tubes. To each test tube containing the insoluble mystery powders, add a few milliliters of iodine tincture. If the unknown turns blue, starch is present. If it remains a brown color (from the iodine tincture), starch is not present. Record your data in the flow chart.
5. Continuing with the insoluble mystery powders that did not contain starch, you will now test for the presence of carbonates by adding vinegar. To the test tubes with insoluble mystery powders that did not contain starch, add a few milliliters of vinegar. If the solution bubbles, the mystery powder contains carbonate. If the solution does not bubble, the mystery powder does not contain carbonate. Record your data in the flow chart.
6. You should now be able to identify three of the nine mystery powders.
7. You will now continue by testing the six water soluble mystery powders for their acid/base character. Carefully dip a piece of red litmus paper into each of the six water soluble mystery powder solutions. Use a separate piece of litmus paper for each solution. If the litmus paper turns blue, the solution is basic. If it does not change color, the solution is acidic. To confirm your results, repeat the test using blue litmus paper. If the blue litmus paper turns pink, the solution is acidic. Record your data in the flow chart.
8. Continuing with the water soluble mystery powders that were basic, you will now add two milliliters of vinegar to test for the presence of carbonate. If the solution bubbles, a carbonate is present. If the solution does not bubble, the mystery powder doesn't contain carbonate. Record your data in the flow chart.
9. You will now perform the same test with the water soluble mystery powders that were acidic. As before, when you add two milliliters of vinegar, if bubbles are generated, the mystery powder contains carbonate. If bubbles are not generated,

the mystery powder does not contain carbonate. Record your data in the flow chart.

10. At this point you will have three remaining mystery powders and you will test for solubility in alcohol. You need to add a pea-sized amount of each of these unknowns to three clean, labeled test tubes. Add two milliliters of 50% rubbing alcohol to each test tube. Record your data in the flow chart.
11. Continuing with the two alcohol-insoluble mystery powders, you will need to add a large pea-sized amount of each mystery powder to two clean, labeled test tubes. The remaining two mystery powders will show slight differences in solubility based on the temperature of the solvent. Add two milliliters of hot water to each test tube. One mystery powder will be more soluble in the hot water than it was in the colder water, one will not. Record your data in the flow chart.
12. Based on your completed flow chart and the properties you researched before lab, you should be able to identify all nine mystery powders.
13. Thoroughly wash all your glassware and tools. Follow the disposal directions of your instructor. Be sure to wash your hands as well.

Cleanup and Disposal

- Cleanup and disposal of all mystery powders and test reagents require no special treatment. You will only be using very small amounts. They can be disposed of in the trash can or down the sink drain with plenty of water. However, follow your instructor's directions.
- Be sure to wash all glassware and tools thoroughly.
- Be sure to wash your hands thoroughly.

Data and Observations

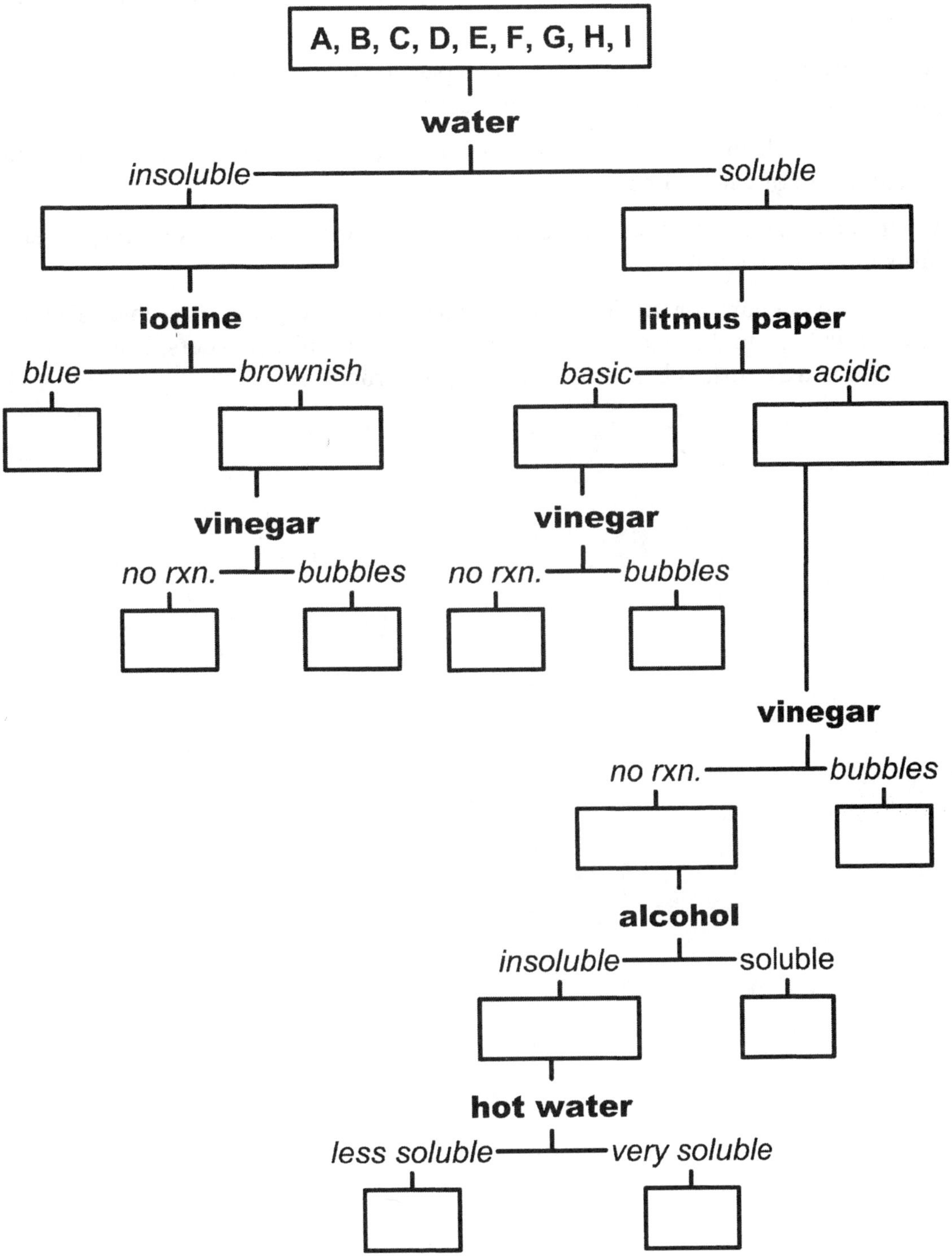

A, B, C, D, E, F, G, H, I

water
- *insoluble* → [ ]
  - **iodine**
    - *blue* → [ ]
    - *brownish* → [ ]
      - **vinegar**
        - *no rxn.* → [ ]
        - *bubbles* → [ ]
- *soluble* → [ ]
  - **litmus paper**
    - *basic* → [ ]
      - **vinegar**
        - *no rxn.* → [ ]
        - *bubbles* → [ ]
    - *acidic* → [ ]
      - **vinegar**
        - *no rxn.* → [ ]
          - **alcohol**
            - *insoluble* → [ ]
              - **hot water**
                - *less soluble* → [ ]
                - *very soluble* → [ ]
            - soluble → [ ]
        - *bubbles* → [ ]

Post Experiment Questions

1. Which of the tests in this experiment tested physical properties and which tested chemical properties?
2. Based upon your experience in this experiment, are chemical properties or physical properties easier to determine? Justify your answer.
3. What is the identity of a substance that is insoluble in water and reacts with vinegar, but not with iodine?
4. What is the identity of a substance that dissolves water, reacts with vinegar, and turns litmus paper red?
5. Design a procedure for the following scenario. You are given a sample of white powder and are told it is a mixture of two different white powders. The possible substances are cornstarch, sucrose and sodium carbonate.

FLAME TESTS

Introduction

In this experiment you will study an important characteristic of atoms—their ability to emit light when they are excited. This characteristic should be familiar if you have ever seen a fireworks display or neon signs. In each of these instances, atoms of a specific substance are excited by an external energy source, and this process produces light of various colors. The color of the light emitted depends on the nature of the atom. Different atoms produce different colors. In other words, the color of the emitted light is like a fingerprint. It can be used to identify the atoms involved.

The color of the light that an excited atom emits also tells us something about the inner structure of that atom. This idea was the basis of the Bohr theory of the atom.

In this experiment you will determine the colors of the light emitted from substances that are excited in the flame of a Bunsen burner.

Background Information

The light emitted by atoms is energy in the form of electromagnetic radiation that we see as visible light. The wavelength of the electromagnetic radiation determines the color of the light we see. Visible light is a small portion of the electromagnetic spectrum, but our eyes can discern minute differences in color. This can be very helpful in determining the approximate wavelength of light that we see emitted in the flame tests. The table below shows wavelength range associated with a particular color. Assigning an exact wavelength to light of a specific color is difficult. As shown in the table, light with a wavelength of 482 nm will appear green, as will light with a wavelength of 510 nm. Light with wavelengths that are in between different colors will look like some combination of those colors. For example, light with a wavelength of 468 nm will appear bluish green or greenish blue.

Color	Wavelength range
Violet	400 – 425 nm
Blue	440 – 460 nm
Green	475 – 520 nm
Yellow	525 – 575 nm
Orange	580 – 615 nm
Red	625 – 750 nm

The light emitted from atoms comes from subatomic particles called **electrons**. The electrons in an atom reside in shells; each successive shell is further and further away

from the nucleus of the atom. Electrons in each shell have a specific energy; electrons in lower shells have a lower energy compared to electrons in higher shells. An atom can be excited by an external energy source. The external energy causes an electron to move from one shell to another shell at a higher energy. When this excited electron falls back to its original shell, it releases energy as a photon of electromagnetic radiation. The energy of the photon released is *inversely* related to the wavelength of the photon. Photons that have more energy have shorter wavelengths and, conversely, photons that have less energy have longer wavelengths.

An important fact that should be noted is that the shells for an element are unique. When an atom of copper is excited and the electron falls from a higher energy shell to a lower one, it releases light at a specific wavelength. All copper atoms will emit light at this wavelength. Different elements will emit light at different wavelengths. Not all elements emit energy as visible light; some elements will only emit very short wavelength photons that we cannot see.

In this experiment we test several solutions containing different metal salts dissolved in water. A Bunsen burner flame will be used to excite the electrons from ions in solution. The excited electrons will fall back to their original state and in the process emit energy. The energy is emitted as photons of electromagnetic radiation and if the energy has a wavelength of between 400 nm and 750 nm we see the energy as visible light. To perform the flame test, we will use a very simple method, but one that may take a little practice.

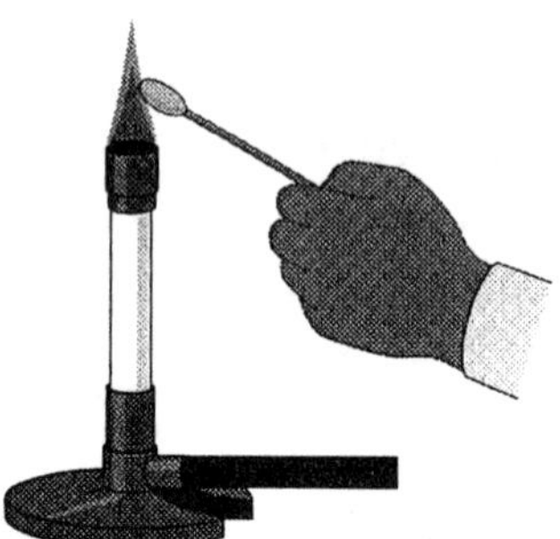

Figure 1. Using a Bunsen burner to perform a flame test.

To perform the flame test, light a Bunsen burner and adjust the flame until it is nearly colorless. Then dip a cotton swab into your test solution and hold the cotton swab near the outside edge of the flame. As the solution evaporates the metal salt will start to emit its characteristic color. CAREFUL: Do not get too close to the flame or you may burn yourself. If you do not see any color after a few minutes, try dipping the swab in the solution again. You do not want the cotton swab to catch on fire!

Chemicals and Equipment

- Bunsen burner
- striker or matches
- cotton swabs
- ~10 mL of 1 M NaCl
- ~10 mL of 1 M KI
- ~10 mL of 1 M NaI
- ~10 mL of 1 M KCl
- ~10 mL of 1 M $CaCl_2$
- ~10 mL of 1 M $CuSO_4$
- ~10 mL of 1 M $BaCl_2$
- ~10 mL of 1 M Borax

Safety Concerns

- **Always wear safety goggles.**
- **Holding the swab over the flame may cause the stem of the cotton swab to melt.**
- **Do not hold the cotton swab over the flame for an extended period of time because the swab itself might ignite.**
- **Keep your hand away from the flame by holding the bottom of the cotton swab.**

Procedure

1. Dip the end of a cotton swab into one of the eight solutions.
2. Hold the cotton swab next to the Bunsen burner flame so that the wet end just touches the flame.

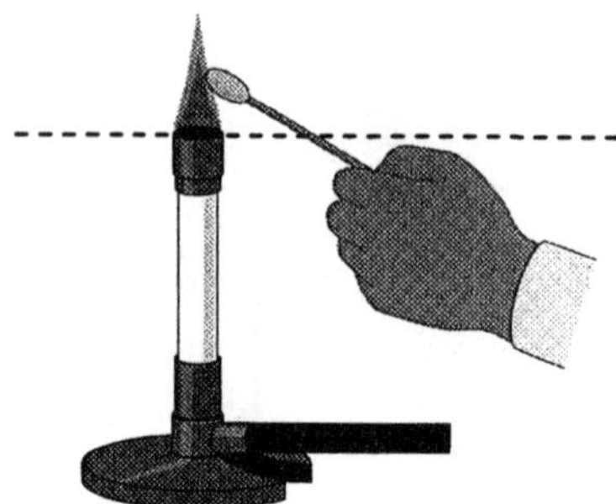

Figure 2. While performing the flame test, keep your hand below the flame to avoid burning yourself!

3. After a few seconds, the characteristic color from the test solution will appear in the flame.
4. Repeat for each of the known solutions.
5. Construct a data table; identify which element can be identified by its characteristic color.

Cleanup and Disposal

☞ Dispose of the leftover solutions in the appropriate containers in your laboratory.

Data and Observations

Data table

Species	Color observed	Element responsible

Post Experiment Questions

1. What conclusions can you make about flame color and the elements in a solution?
2. Which elements are responsible for the different colors you observed?
3. If you saw a blue flame, would that have more energy than a yellow flame?
4. What color would you expect to see in a flame test for a solution made by combining sodium chloride and copper sulfate?

Information for Instructors

1 M NaCl: add 14.6 g NaCl to 250 mL of distilled or deionized water
1 M KI: add 415 g KI to 250 mL of distilled or deionized water
1 M NaI: add 37.5 g NaI to 250 mL of distilled or deionized water
1 M KCl: add 18.6 g KCl to 250 mL of distilled or deionized water
1 M $CaCl_2$: add 35.2 g $CaCl_2$ to 250 mL of distilled or deionized water
1 M $CuSO_4$: add 40 g $CuSO_4$ to 250 mL of distilled or deionized water
1 M $BaCl_2$: add 52 g $BaCl_2$ to 250 mL of distilled or deionized water
1 M Borax: add 95.3 g Borax (sodium tetraborate decahydrate, $Na_2B_4O_7 \bullet 10H_2O$) to 250 mL of distilled or deionized water

Combustion of Hydrocarbons and Alcohols

Introduction

In this experiment you will study the heat released during the combustion of different fuels. The fuels studied in this experiment are hydrocarbons and alcohols, which are hydrocarbons that contain oxygen. Combustion is the reaction of a fuel with oxygen; the reaction releases heat and produces, among other things, carbon dioxide and water. Measuring the change in temperature of water in a container suspended above the flame will provide the data to determine the amount of heat released.

Background Information

Hydrocarbon fuels are produced from decomposed organic material in Earth's surface, such as oil, natural gas, and coal. Refining these natural resources produces the fuels used in homes, cars, and power plants. Chemically, hydrocarbons are molecules containing carbon and hydrogen. They are found in long chains, branched chains, or cyclical arrangements. When reacted with oxygen, hydrocarbons produce carbon dioxide, carbon monoxide, water, and other products. The length of the carbon chains determines the properties of the compound. Shorter carbons like methane, CH_4, and propane, C_3H_8, are gases under normal conditions. Hydrocarbon chains containing six or eight carbons atoms, like hexane, C_6H_{14}, and octane, C_8H_{18}, are liquids. Gasoline is a mixture of different hydrocarbons that contain about eight carbon atoms. Oils used in lamps are another example of hydrocarbons that contain about twelve carbon atoms. Longer hydrocarbon chains that contain 20, 30, 40, or more carbon atoms are solids at room temperature. Candle wax is a long-chain hydrocarbon that contains about 40 carbon atoms.

Hydrocarbons that include a hydroxide group (–OH) are called alcohols. The alcohol used in beverages is called ethanol and has a chemical formula of C_2H_5OH. The combustion of alcohols, like the combustion of hydrocarbons, produces carbon dioxide and water, and releases enough heat to make alcohols a useful fuel source. Sterno® contains a mixture of methanol and other alcohols combined with a jelly in a can; it is commonly used in the food service industry.

A property of a fuel is the amount of heat it releases during combustion. Heat is a type of energy measured in calories. The amount of heat released is different for each fuel. In this lab you will have the option of either measuring several times the heat released by one type of fuel or measuring the heat released by different fuels. In either case you will have to make careful mass and temperature measurements to calculate the heat contained by different fuels.

Chemicals and Equipment

- aluminum can
- scissors
- ruler
- glass rod
- alcohol lamps containing different fuels.
- matches
- thermometer
- ring stand with iron ring
- 100-mL graduated cylinder
- balance (capable of measuring 0.001 g)

Safety Concerns

- **Always wear safety goggles.**
- **You will be working with flammable solvents in the presence of open flames. Use caution to avoid spilling any solvents and handle alcohol lamps with extreme caution. Before you start the experiment, your instructor should inform you what to do in case of a spill or fire.**
- **Make sure loose clothing and long hair are tied back so they do not come in contact with open flames.**
- **Use tongs or heat-resistant gloves to handle hot materials.**

Procedure

1. Construct an aluminum water vessel by removing the top of a soda can with a pair of scissors. Cut the soda can so the height of the vessel is 9–9.5 cm. Be careful not to cut yourself! To complete the water vessel, poke two holes through opposite sides of the soda can, about 2-cm from the top. The holes should be large enough for a glass rod. When complete, make sure the can is dry and slide a glass rod through the holes.
2. Adjust the height of the ring and the thermometer clamp on the ring stand so that the bottom of the can is 2.5-cm above the top of the wick of the alcohol lamp.

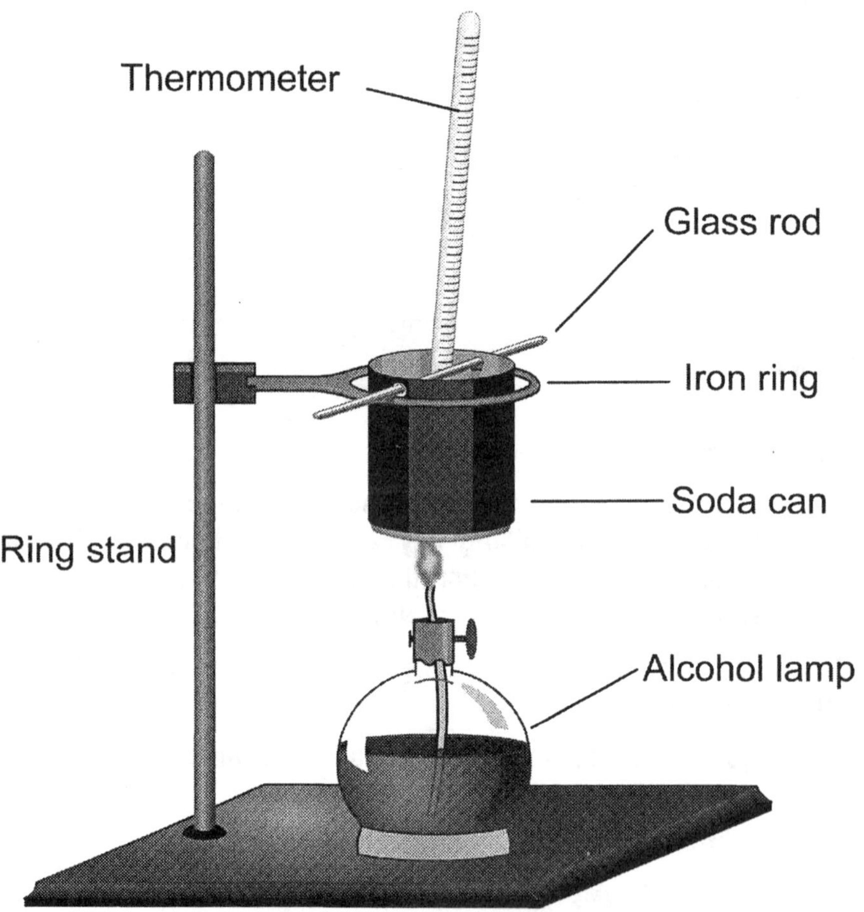

Figure 1. Experimental setup

3. Weigh the empty can. Make sure the balance reads 0.0 g before placing your can on the balance. Record the mass to the nearest 0.1 g.
4. Using a graduated cylinder, add 100-mL of water to the can. Weigh the can with water. Record the mass to the nearest 0.1 g.

5. Measure the temperature of the water in the can. Stir the water for 10 to 15 seconds and record the temperature to the nearest 0.1°C (you may have to estimate the tenths of a degree if the thermometer has markings every degree).
6. Obtain an alcohol lamp from your instructor. Record the name of the fuel in the data table. Weigh the alcohol lamp. Make sure the balance reads 0.00 g or 0.000 g before placing the alcohol lamp on the balance. Record the mass to the nearest 0.01 g or 0.001 g.
7. Place the alcohol lamp under the can and light the lamp. You may have to adjust the wick of the alcohol lamp so that the flame is just below the can.
8. Continue to heat the water, stirring the water occasionally, until the temperature has increased about 25°C. Extinguish the flame.
9. Immediately after extinguishing the flame, stir the water *gently* until the temperature stops rising. Record the temperature to the nearest 0.1°C.
10. Weigh the alcohol lamp. As in Step 6, make sure the balance reads 0.00 g or 0.000g before placing the alcohol lamp on the balance. Record the mass to the nearest 0.01 g or 0.001 g.
11. Before performing another trial, discard the water in the can. Thoroughly clean and dry the can with water and a paper towel; *be careful not to cut yourself on the top of the can.*

Calculations

The following values will have to be calculated from the data collected during the experiment:

- the mass of the water
- the temperature change of the water
- the mass of the fuel used
- the heat absorbed by the water
- the heat produced per gram of fuel

The mass of the water used in the experiment is calculated by subtracting the mass of the empty can from the mass of the filled can:

Mass of water (g) = mass of can and water (g) – mass of can (g)

In the data table, the entry for the mass of the empty can is below the entry for the mass of the water-filled can. Setting up the data table this way makes it easier to calculate the mass of the water used in the experiment. The temperature change and the mass of fuel used in the experiment are also calculated by subtraction.

To calculate the heat absorbed by the water, multiply the mass of the water, the temperature change of the water, and the specific heat of water:

Heat absorbed (cal) = mass of water (g) × temperature change (°C) × 1.00 cal/g•°C

The specific heat of a substance relates the amount of energy it takes to raise one gram of the substance by one degree Celsius. The energy can be in calories, joules, or some other unit of energy. For simplicity, we use calories in this experiment. It takes one calorie of energy to raise one gram of water by one degree Celsius.

To calculate the amount of heat released by a fuel, divide the heat absorbed by the water by the mass of fuel used:

Heat released per gram (cal/g) = heat absorbed (cal) ÷ mass of fuel used (g)

Data and Observations

Data Table

Trial number	1	2	3
Name of Fuel			
Mass of can and water (g)			
Mass of can (g)			
Mass of water (g)			
Initial mass of lamp (g)			
Final mass of lamp (g)			
Mass of fuel burned (g)			
Final temperature of water (°C)			
Initial temperature of water (°C)			
Temperature change (°C)			
Total heat absorbed by water (calories)			
Heat per gram of fuel (calories/gram)			

Post Experiment Questions

1. Determine the average heat released per gram of fuel for the different fuels used in you and your classmates. This may require sharing data with other students.
2. Create a table that ranks the different fuels by the heat released per gram of fuel; include the chemical formula in your table.
3. Make a general rule about the heat released per gram of fuel and the chemical formula of the fuel. Give a reasonable explanation for the rule.
4. How does the oxygen content affect the heat released by the fuel?
5. Carbohydrates, or sugars, have an empirical formula of CH_2O, or a little more than 50% oxygen by mass. Would you expect carbohydrates to release more or less energy per gram than fats, which are typically 10% oxygen by mass? Why?

Information for Instructors

Fill the different alcohol lamps with the following:
Methanol, CH_3OH
Ethanol, C_2H_5OH
2–Propanol (rubbing alcohol), C_3H_7OH

Comparisons to hydrocarbon fuels can be made by using lamp oil or candle wax as additional fuel sources. Lamp oil is typically a mixture of hydrocarbons, but can be approximated as $C_{12}H_{26}$. Candle wax is also a mixture of hydrocarbons, and it can be approximated as $C_{40}H_{82}$.

MAKING YOUR OWN FUEL: BIODIESEL

Introduction

In this experiment, you will prepare biodiesel fuel from vegetable oil. Biodiesel is clean burning alternative fuel made from vegetable oil, which is a renewable resource. The conversion of vegetable oil to biodiesel fuel is carried out through a process called esterification.

Background Information

Rudolf Diesel patented the diesel engine in 1892, as a more efficient alternative to the gasoline engine. Both diesel and gasoline engines burn hydrocarbon fuels to produce energy. Hydrocarbons are chains of carbon atoms with hydrogen atoms bonded to each carbon atom. The length of the carbon-atom chain is used to describe different hydrocarbons. Gasoline is a mixture of hydrocarbons with eight and nine carbon-atom chains. Diesel fuel is a mixture of longer-chain hydrocarbons, consisting of 12, 14, or 16 carbon atoms. Because the two fuels have different properties, gasoline engines cannot use diesel fuel and diesel engines cannot use gasoline. The advantage of using a diesel engine is its fuel efficiency, or gas mileage. However, both gasoline and diesel fuel are derived from crude oil, a nonrenewable resource. An alternative to petroleum-based diesel fuel is biologically derived diesel fuel, or biodiesel, produced from vegetable oil. Biodiesel has been officially recognized by the Department of Energy and the Department of Transportation as an alternative fuel, and it meets the requirements of the Clean Air Act. Biodiesel can be used on its own as a fuel or blended with regular petroleum-based diesel fuel. A 20% blend of biodiesel with regular diesel fuel is called B20 diesel fuel.

Biodiesel is produced from vegetable oil by reacting the oil with an alcohol in the presence of lye. The oil is converted into an ester, which contains oxygen atoms within the hydrocarbon chain. The incorporation of the oxygen atoms in biodiesel fuel causes it to burn cleaner than normal petroleum-based diesel fuels. The production of biodiesel is carried out in two steps, shown in Figure 1. First, potassium hydroxide, KOH, is added to methanol. The hydroxide reacts with the methanol to produce methoxide ion and water. The second step involves adding vegetable oil to the methanol–potassium hydroxide mixture. Three methoxide ions react with a molecule of vegetable oil to produce three methyl ester molecules and one glycerol molecule. The long-chain methyl esters are biodiesel fuel and form a separate layer from the glycerol, making it easy to separate the two products.

The procedure used in this experiment will produce actual biodiesel fuel; however, before it is suitable to use in an automobile, the biodiesel would need to be purified further to remove any unreacted methanol or water that may have been produced.

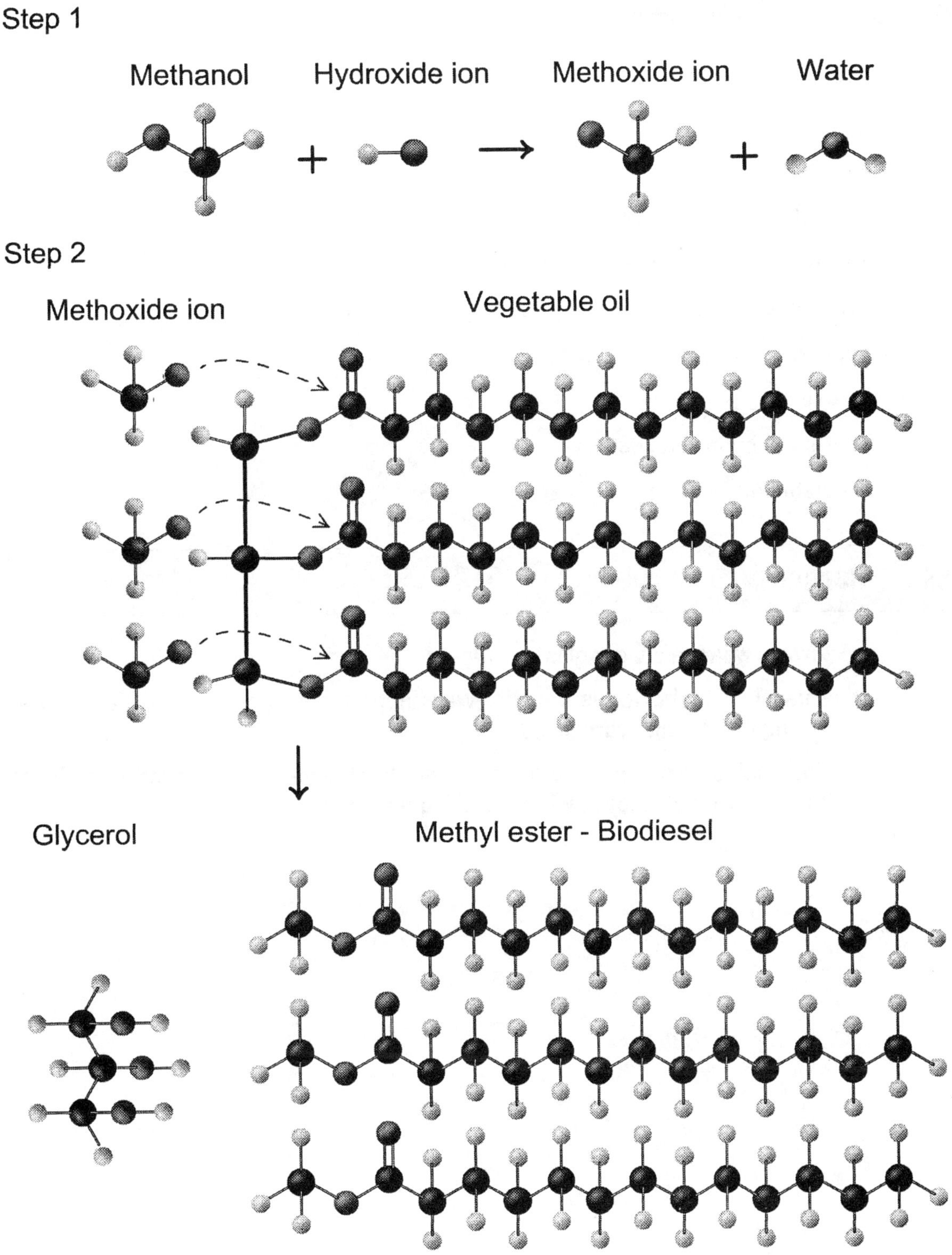

Step 1

Methanol   Hydroxide ion   Methoxide ion   Water

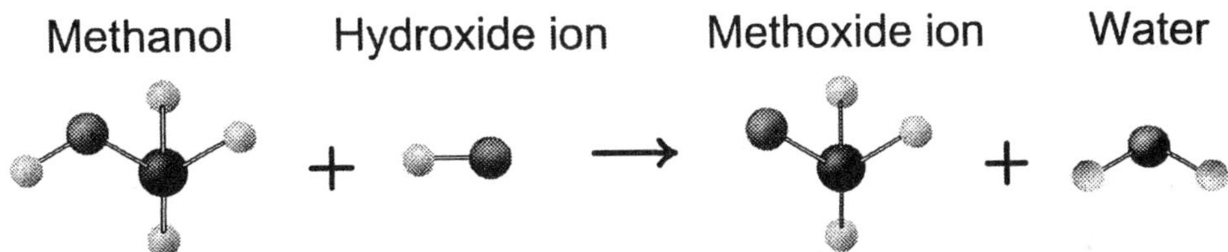

Step 2

Methoxide ion

Vegetable oil

Glycerol

Methyl ester - Biodiesel

Figure 1. A ball-and-stick model of biodiesel production from vegetable oil.

Chemicals and Equipment

- two 250-mL beakers
- glass rod
- thermometer
- 100-mL graduated cylinder
- 25-mL or 50-mL graduated cylinder
- balance (capable of measuring 0.001 g)
- hot plate
- methanol
- potassium hydroxide pellets
- vegetable oil

Safety Concerns

- **Always wear safety goggles.**
- **Potassium hydroxide is caustic. Wear appropriate gloves when handling and mixing potassium hydroxide.**
- **Methanol is flammable and toxic. Do not handle methanol around open flames. Wear appropriate gloves when handling methanol.**

Procedure

1. Using a graduated cylinder, add 100-mL of vegetable oil to the 250-mL beaker and heat to 50°C on a hot plate. Complete steps 2–4 while the vegetable oil is heating.
2. Weigh 1.2 g of KOH pellets in a weighing vessel.
3. Using a clean and dry graduated cylinder, add 25-mL of methanol to a second 250-mL beaker. Make sure the beaker is totally clean and dry. (If water is present, you will end up making *soap*, not biodiesel!)
4. SLOWLY add the KOH pellets to the methanol while stirring the solution. Use caution not to splash the solution on your skin. If you get any solution on your skin, wash it off immediately with cold water. Gently stir the solution until all of the KOH pellets have dissolved.
5. Remove the warm vegetable oil from the hot plate. Slowly add the warm vegetable oil to the KOH–methanol solution while stirring briskly. After all of the vegetable oil has been added, place the beaker on a hot plate and maintain a temperature of between 45°C and 50°C while stirring vigorously for 15 minutes. No discernable layers should be present afterward.
6. Remove the beaker from the hot plate and allow it to cool. As the mixture cools, two layers will form. The top layer is biodiesel; the bottom layer is mostly glycerol.
7. Carefully separate the bio-diesel from the bottom layer by pouring the biodiesel into a graduated cylinder. Record the volume of biodiesel fuel produced.
8. When complete, discard the products into the appropriate waste containers. Clean your glassware with warm soapy water.

Data and Observations

Original volume of vegetable oil: ___________

Volume of biodiesel fuel produced: _____________

Data table

Substance	Description
Vegetable oil	
Biodiesel fuel	
Glycerol	

Post Experiment Questions

1. Using the volumes of oil and fuel, what is the percentage of biodiesel conversion in this experiment?
2. If you wanted to fill a 40 L gas tank (about 10.5 gallons) with biodiesel fuel, how much vegetable would you have to use?
3. Compare the viscosity—how easily the liquid flows—of biodiesel to vegetable oil. Which would be easier to pump through a vehicle's fuel system?
4. If possible, cool a small amount of biodiesel fuel and vegetable oil in an ice bath. How does the temperature affect the viscosity? What would be the impact on using biodiesel or vegetable oil for automotive fuels?
5. In a previous experiment, you measured the temperature change of water caused by burning hydrocarbon fuels. If time permits, repeat that experiment using biodiesel in the alcohol burner. You may have to team up with a classmate to have an adequate supply of biodiesel.

PLASTICS: ANALYSIS, CLASSIFICATION, AND IDENTIFICATION

Introduction

The goal of this experiment is to be able determine the identity of an unknown plastic based upon its physical and chemical properties. The unknown plastic is one of the six most common commercial plastics. Six tests will be carried out on a sample of each type of plastic and the unknown sample. Three of the tests involve testing the density of the plastics in different solvents—water, isopropyl alcohol, and corn oil. The other tests are testing the color emitted by the plastic when it is burned, the solubility of the plastic in acetone, and whether the plastic softens in hot water. Upon determining the properties of the known samples of plastics, you will be able to identify the unknown plastic.

Background Information

Think of everything you have done today. How many times have you used a plastic? Your alarm clock may have been made of plastic, your toothbrush is plastic, the orange juice container and milk container were made of plastic, and even your car contains many plastic parts. Even though you use and encounter plastics on a daily basis, it is unlikely that you are familiar with or understand their classification system. There are more than 60,000 different types of plastics manufactured. However, six of those types account for 70% of the plastic used in the United States. The six plastics are polyethylene terephthalate (PET), high-density polyethylene (HDPE), polyvinyl chloride (PVC), low-density polyethylene (LDPE), polypropylene (PP), and polystyrene (PS).

To help identify the six plastics, they are coded with a triangle of chasing arrows and a number inside the triangle. These codes are used when recycling plastics.

Figure 1. Recycling code found on the bottom polypropylene containers.

The codes for the six plastics are as follows:

Code number	Name	Abbreviation	Example
1	Polyethylene terephthalate	PET	Soft-drink bottle
2	High-density polyethylene	HDPE	Milk container
3	Polyvinyl chloride	PVC	Cooking oil bottle
4	Low-density polyethylene	LDPE	Elmer's glue bottle
5	Polypropylene	PP	Yogurt container
6	Polystyrene	PS	Deli container

You will perform six tests that show differences in the properties of the various plastic samples. You will first examine the density of each plastic in relation to the density of water. If the sample floats, it is less dense than water. However, if it sinks, it is more dense than water. The isopropyl alcohol test will examine the density of the plastic samples in relation to the density of isopropyl alcohol. You will do the same with corn oil. You will also perform a flame test, a heat test in boiling water, and a solubility test using acetone.

Because labeling is voluntary, plastics that have not been labeled with the code may be difficult to identify. You will receive an unknown sample on which you will perform the same six tests. Based on the results of the tests, as well as, the sample's general appearance, you will be able to identify this unknown plastic sample.

Chemicals and Equipment

- 6 identified plastics—you will need 2 or 3 pieces of each sample
- 1 unknown plastic—you will need 2 or 3 pieces of this sample
- 250-mL beaker
- Bunsen burner
- pairs of tongs
- glass stirring rod
- 6 pieces of copper wire, ~5 cm each
- ~200 mL tap water

Safety Concerns

- **Four of the tests have been set up in the laboratory room as stations. You will perform these tests at the stations rather than at your laboratory station due to the flammability of the chemicals.**
- **Acetone is highly flammable and must always be kept away from flames and the hot plate. It must be kept covered when not in use. You will use the glass petri dish as a cover for the 100-mL beaker.**
- **Isopropyl alcohol also is flammable and must be kept away from flames and the hot plate.**
- **Corn oil is also flammable and must be kept away from flames and the hot plate.**
- **The heat test using boiling water to determine the softening point uses a hot plate that must be kept away from the flammable liquids and is set up as a station.**
- **The two remaining tests, the water test and the copper-wire flame test, will be performed at your laboratory station.**
- **Safety goggles should be worn at all times.**

Procedure

Water test

This test is performed at your laboratory station.

1. Place the plastic sample in a 250-mL beaker to which you have added approximately 100 mL of tap water at room temperature.
2. Using a glass stirring rod, dislodge any air bubbles that may have adhered to the plastic and then try to sink the sample.
3. Some of the samples will sink and some will float. Record your observations on the data chart.
4. You should remove the plastic sample and dry it off. You will continue to use the samples in other tests.
5. Repeat this procedure for each plastic sample.

Copper-wire test

This test is performed at your laboratory station.

6. Obtain a piece of copper wire, about 5 cm in length, and using tongs hold it in the hottest part of a Bunsen burner flame until it is red hot.
7. Be careful not to touch the red-hot copper wire as you remove it from the flame and touch it to the plastic sample.
8. If the plastic sticks to the hot copper wire, carefully pry it off with another pair of tongs.
9. Save the remainder of the plastic sample for more tests.
10. The hot copper wire should now have a glob of plastic stuck to it. Place the copper wire with the plastic glob back into the Bunsen burner flame.
11. You should notice either a green flame or an orange flame. Record your observations on the data chart.
12. Quickly remove the copper wire with the plastic glob and quench the burning/hot sample in a beaker of water to stop the burning of the plastic and to cool the wire.
13. Place the copper wire with the plastic glob in the appropriate waste container.
14. Repeat this procedure for each plastic sample.

Acetone test

This test is not performed at your laboratory station. A special station has been set up. Acetone is highly flammable and must be kept away from flames and the hot plate. Be sure to cover the acetone with the glass petri dish when you are finished.

15. Using tongs, place the plastic sample for about 20 seconds in a beaker that contains acetone.
16. Remove the sample and press it between your fingers. If the acetone has "loosened up" the polymer chains of the plastic, it will feel soft and sticky.
17. Then try to scrape off some of the plastic with a fingernail. You can scrape off some of the plastic for some of your samples, but not for all of them.
18. Be sure to replace the petri dish on top of the beaker.
19. Record your observations in the data chart.
20. After performing this test, discard the plastic samples in the appropriate waste containers and obtain new plastic samples.
21. Repeat this procedure for each plastic sample.

Heat test

This test is not performed at your laboratory station. A special station has been set up. The hot plate must be kept away from the flammable chemicals.

22. Using tongs, hold the plastic sample in a beaker of boiling water for 30 seconds.
23. Although the samples will not melt, some may soften at 100°C. Remove the sample from the water and press it between your fingers to see if it has softened.
24. Record your observations on the data chart.
25. If the sample has softened and distorted you may wish to obtain fresh samples; otherwise, keep the samples for further tests. If you need to dispose of the samples, place them in the appropriate waste containers.
26. Repeat this procedure for each plastic sample.

Isopropyl alcohol test

This test is not performed at your laboratory station. A special station has been set up. Isopropyl alcohol is flammable and must be kept away from flames and the hot plate.

27. Place the plastic sample in a beaker that contains isopropyl alcohol.
28. Using a glass stirring rod, dislodge any air bubbles that may have adhered to the plastic and then try to sink the sample.
29. Some of the samples will sink and some will float. Record your observations on the data chart.
30. You should remove the plastic sample with a plastic spoon and dry it off. You will continue to use the samples in other tests.
31. Repeat this procedure for each plastic sample.

Corn oil test

This test is not performed at your laboratory station. A special station has been set up. Corn oil is flammable and must be kept away from flames and the hot plate.

32. Place the plastic sample in the beaker that contains corn oil.
33. Using a glass stirring rod, dislodge any air bubbles that may have adhered to the plastic and then try to sink the sample in the cooking oil.
34. Some of the samples will sink and some will float. Record your observations on the data chart.
35. You should remove the plastic sample with a spoon and clean it off.
36. Repeat this procedure for each plastic sample.
37. Since this is the last test, place your plastic samples in the appropriate waste containers. If this was not your last test, keep the samples and continue the experiment.

Cleanup and Disposal

- All plastics samples should be placed in the appropriate waste containers.
- The copper wire should be placed in its appropriate waste container.
- The cooking oil may be either washed down the sink or placed in a waste container. Follow your instructor's directions.
- The acetone should be placed in its appropriate waste container.
- The isopropyl alcohol should be placed in its appropriate waste container.
- Be sure to wash thoroughly all glassware and equipment.
- Be sure to wash your hands thoroughly.

Data and Observations

Test	PET	HDPE	PVC	LDPE	PP	PS	Unknown
General appearance							
Water							
Copper wire							
Acetone							
Heat							
Isopropyl alcohol							
Corn oil							

Post Experiment Questions

1. Based on your results, what is the identity of your unknown sample? How did you come to this conclusion?
2. You performed all six tests on all of your samples. This was time-consuming. Based on your observations, construct a flow chart that would enable you to perform as few tests on as few samples as possible. Hint: begin with the water test like the chart below.

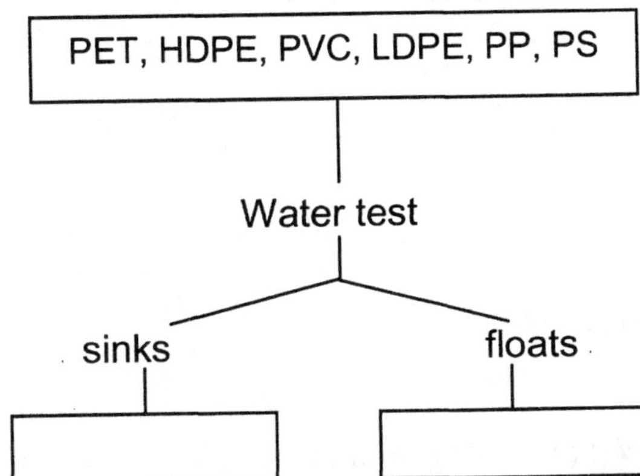

Information for Instructors

Four separate test stations must be prepared in order to keep flammable materials from open flames or hot plates. The stations should have the following equipment:

Acetone test

- 250-mL beaker
- glass petri dish—used to cover 250-mL beaker
- ~100 mL acetone
- pair of tongs

Isopropyl alcohol test

- 250-mL beaker
- ~100 mL isopropyl alcohol
- glass stirring rod
- plastic spoon—used to retrieve sunken samples

Heat test

- hot plate
- pair of tongs
- 250-mL beaker
- ~100 mL boiling water

Corn oil test

- 250-mL beaker
- ~100 mL isopropyl alcohol
- glass stirring rod
- plastic spoon—used to retrieve sunken samples
- ~100 mL corn oil *

The Detection of Alcohol in Breath

Introduction

The goal of this experiment is to detect alcohol in a system similar to a breathalyzer test. Air is blown through an alcohol containing solution and then into a potassium dichromate solution. Based upon the amount of time it takes for the potassium dichromate to change colors, the amount of alcohol present can be determined.

Background Information

Alcohol is one of the most used and most misused substances in our country. Using and abusing alcohol affects us in many ways. Alcohol is the opposite of a stimulant. It actually slows down your reflexes, making it especially dangerous for you to be behind the wheel after consuming alcohol. All states have legal limits for blood alcohol concentration, and police officers can conduct several tests to measure blood alcohol levels.

Although commonly called "alcohol", it is more accurate to refer to the alcohol found in beverages as ethanol or ethyl alcohol since there are many other types of alcohols in use. In this experiment you will simulate a system that detects ethanol concentration. You will be using a solution that contains dichromate ion, $Cr_2O_7^{2-}$, which is a yellow-orange color. The dichromate ion will oxidize ethanol into a carboxylic acid while it is being reduced to a Cr^{3+} ion, which is green in color. Depending on the ethanol concentration present, the time it takes for this reaction to occur will vary.

$$3\,CH_3CH_2OH + \underset{\text{yellow-orange}}{2\,Cr_2O_7^{2-}} + 16\,H^+ \rightarrow 3\,CH_3COOH + \underset{\text{green}}{4\,Cr^{3+}} + 11\,H_2O$$

Many of the chemicals that you will use in this experiment are caustic or toxic. It is imperative that you listen carefully to your instructor's directions concerning the handling and disposal of all chemicals.

Chemicals and Equipment

- 30.0 g potassium dichromate, $K_2Cr_2O_7$
- 30.0 mL concentrated (18 M) sulfuric acid, H_2SO_4
- ~150 mL absolute ethanol
- deionized or distilled water
- unknown (alcohol concentration) sample
- balance
- weighing paper or weighing boat
- 600-mL Erlenmeyer flask with 2-holed stopper
- 250-mL or 125-mL Erlenmeyer flask
- 100-mL beaker
- 500-mL beaker
- 100-mL graduated cylinder
- graduated disposable pipet
- 4 pieces of glass tubing (see experiment setup)
- 2 pieces of rubber tubing (see experimental setup)
- Ring stand and clamp

Safety Concerns

- **Potassium dichromate is toxic and is human carcinogen. Use caution when handling it. If you get any on your skin, flush with plenty of cold water. Cleanup any spills immediately.**
- **Concentrated sulfuric acid should be used in a hood to prevent inhalation of fumes. Use caution when handling it. If you get any on your skin, flush with plenty of cold water. Cleanup any spills immediately.**
- **Because of the caustic and toxic nature of the chemicals used in this experiment, listen to your instructor's directions carefully concerning the handling and disposal of all chemicals.**
- **Goggles should be worn at all times.**

Procedure

Preparation of dichromate solution

1. Obtain a 600-mL beaker and add 400 mL of water by using a 100-mL graduated cylinder. This beaker should be labeled "dichromate solution."
2. Using a balance and weighing paper, obtain 38.0 g of potassium dichromate, $K_2Cr_2O_7$, and transfer it to the beaker.

At this point, you should work in a hood.

3. To this beaker, carefully add 40.0 mL of concentrated (18 M) sulfuric acid, H_2SO_4. You should measure the volume of acid by using a 100-mL graduated cylinder.
4. Carefully swirl until all of the components are dissolved. Be careful, since this process will generate heat.
5. At the end of the experiment, any leftover dichromate solution should be placed in the appropriate waste container. Listen carefully to your instructor's directions concerning the disposal of this solution.

Developing a control sample for ethanol content

6. To prepare your control sample, measure 50.0 mL of the dichromate solution by using a graduated cylinder, and add it to a 100-mL beaker that has been labeled "control sample."
7. To this beaker, add 0.1 mL of ethanol using a graduated disposable pipet.
8. Record the resulting color in the data table. Set the beaker aside to use as a reference for future tests.

Based on the following illustration, gather the necessary equipment and set up your ethanol detection system.

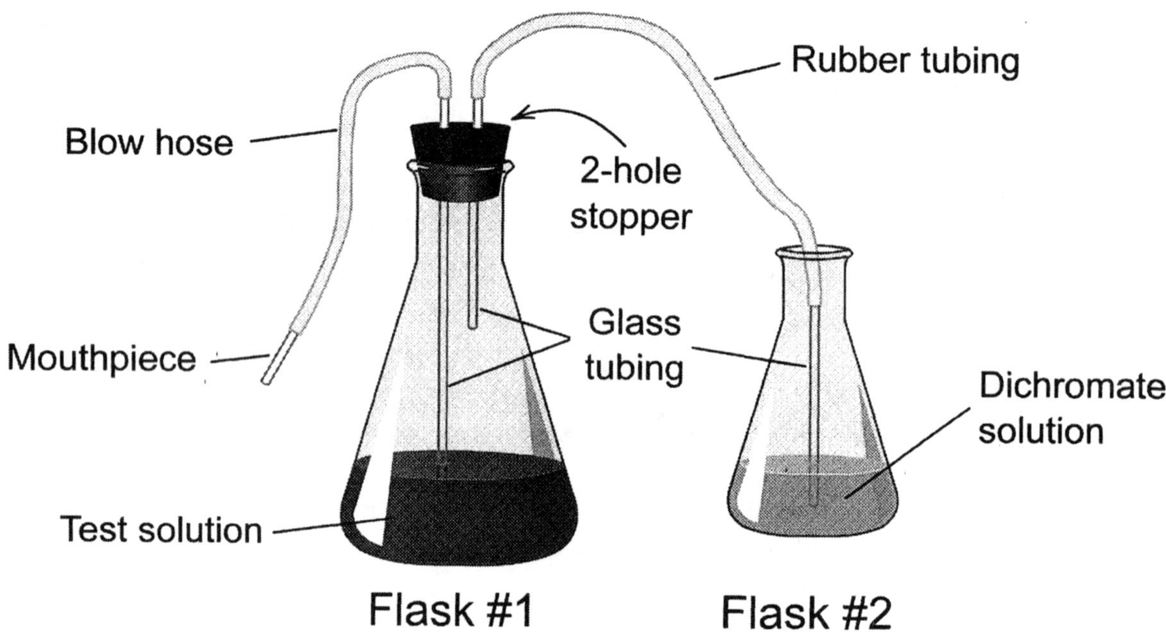

Figure 1. Ethanol detection setup.

Developing a standard curve for ethanol content

9. Using a graduated cylinder, add 50.0 mL of the dichromate solution to flask #2. Flask #2 may be either a 250- or 125-mL Erlenmeyer flask. Secure this flask by placing it on a ring stand with a clamp around the neck of the flask.
10. Place flask #2 back into your setup, making sure that the glass tube is immersed in the dichromate solution.
11. You will be changing the solutions in flask #1. Flask #1 should be a 500-mL Erlenmeyer flask.
12. Measure 50.0 mL of ethanol using a graduated cylinder and transfer it to flask #1.
13. Measure 50.0 mL of water and add it to the ethanol in flask #1. You have now created a 50% ethanol solution. All data that results from this solution should be recorded in the appropriate location.
14. Place flask #1 back into your setup, making sure that the longer glass tube (the one into which your blow) is immersed in the ethanol solution.
15. Blow into the glass mouthpiece for 60 seconds. Be sure to use long even breaths and do not breathe in when your mouth is on the glass mouthpiece.

16. As soon as you stop blowing, begin timing the amount of time that it takes for the dichromate solution in flask #2 to turn the same color as the "control sample." Record the time in the data table.

17. The solution from flask #2 should be transferred to the appropriate waste container. Listen to your instructor's directions carefully concerning the disposal of your waste. After disposing of the dichromate solution, thoroughly wash flask #2.

18. The solution from flask #1 can be disposed of in a sink. Thoroughly wash flask #1.

Repeat steps 9–18 using a different alcohol test solution. The following alcohol solutions should be tested:

- 40% ethanol solution—mix 40 mL of ethanol and 60 mL of water
- 30% ethanol solution—mix 30 mL of ethanol and 70 mL of water
- 20% ethanol solution—mix 20 mL of ethanol and 80 mL of water
- 10% ethanol solution—mix 10 mL of ethanol and 90 mL of water

Determining the ethanol content in the unknown sample

19. Obtain a sample of ethanol with unknown concentration.

20. Record the sample's identification number in the data table.

21. Transfer the unknown solution to flask #1 and return the flask to your setup, making sure that the longer glass tube (the one into which your blow) is immersed in the ethanol solution.

22. Using a graduated cylinder, add 50.0 mL of the dichromate solution to flask #2 and return the flask to your set up, making sure that the glass tube is immersed in the dichromate solution.

23. Blow into the glass mouthpiece for 60 seconds. Be sure to use long even breaths and do not breathe in when your mouth is on the glass mouthpiece.

24. As soon as you stop blowing, begin timing the amount of time that it takes for the dichromate solution in flask #2 to turn the same color as the "control sample." Record the time in the data table.

25. The solution from flask #2 should be transferred to the appropriate waste container. Listen to your instructor's directions carefully concerning the disposal of your waste. After disposing of the dichromate solution, thoroughly wash flask #2.

26. The solution from flask #1 can be disposed of in a sink. Thoroughly wash flask #1.

Cleanup and Disposal

- Be sure to wash thoroughly all of the equipment and supplies used during this lab. Be sure to wash your hands thoroughly at the end of this lab.
- Many of the chemicals that you have used in this experiment are caustic or toxic. It is imperative that you listen carefully to your instructor's directions concerning the handling and disposal of all chemicals.

Data and Observations

General observations:

Control sample color: ____________________

Data table

Experimental solution	Composition	Time to change color (minutes)
50%	(50 mL ethanol and 50 mL water)	
40%	(40 mL ethanol and 60 mL water)	
30%	(30 mL ethanol and 70 mL water)	
20%	(20 mL ethanol and 80 mL water)	
10%	(10 mL ethanol and 90 mL water)	
Unknown		

Unknown sample concentration: ____________________ %

Unknown sample identification #: ____________________

A common method for reporting results is to make a graph. You are to graph your results with the independent variable being placed on the X or horizontal axis and the dependent variable being placed on the Y or vertical axis. The first step is to decide if the time or the solution concentration is the independent variable. The independent variable is defined as the variable that you directly manipulate. The dependent variable is defined as the variable that responds to the independent variable.

Independent variable: ____________________

Dependent variable: ____________________

Make sure that you include the following items on your graph:

- Title of graph
- Each axis properly labeled including units
  (For example: Alcohol Concentration (%))
- Numerical increments on each axis
  The numbers that you use will be determined by your data. You want your graph to fill most of the graph paper and have scale divisions on each axis that are easy to read and use. For example 0.5 min, 1 min., 1.5 min, etc., but not 2/3 min, 1 1/3 min, 2 min, etc.
- Use a ruler to draw the line of best fit through your data points to help illustrate the trend that your data conveys.
- Use your graph to determine the alcohol concentration in your unknown sample. Find the time to change color on your graph axis. Draw a line from the time on the graph axis to your line-of-best-fit. Then determine the alcohol concentration at that point from the concentration axis on your graph.

Post Experiment Questions

1. Describe the process of how the ethanol reacts with the dichromate ion in the system used in this experiment.
2. Why is it necessary to have a control sample?
3. What is the importance of blowing into the mouthpiece with long even breaths?
4. Did your data produce a positive correlation or a negative correlation? What does this imply? From your graph evaluate how dependable is the correlation of time with alcohol concentration?
5. Would this be a practical test for police officers to carry in their squad cars? What led you to this answer? If you needed to make modifications, what would they be?

Determining the Vitamin C Content in Common Substances

Introduction

The purpose of this experiment is to determine the vitamin C content in various products from the grocery store. You will use an iodometric titration method to determine the vitamin C content of a sample with a known amount of ascorbic acid, a store-bought vitamin C tablet, and an orange juice sample.

Background Information

Vitamin C, or ascorbic acid ($C_6H_8O_6$), is a common and essential vitamin that our bodies cannot synthesize and that therefore we must acquire through diet or supplements. The 2002 U.S. RDA (United States recommended daily allowance) for an average woman is 75 milligrams per day, and 90 milligrams per day for an average man. Vitamin C tablets are sold in pharmacy stores as a supplement to a normal diet. Besides tablets, natural sources provide an ample supply of vitamin C to meet our bodys' need. Vitamin C is a common component of citrus fruits, tomatoes, white potatoes, berries, and many other fruits and vegetables.

As cold season approaches, many people increase their vitamin C intake. When you are fighting a cold or an infection, your body increases its vitamin C absorption. Scurvy, a disease formerly common on ships, is a disease characterized by loose teeth and hemorrhaging, and is due to a vitamin C deficiency. Small amounts of vitamin C can eliminate scurvy. Ships would carry lemons or limes on long trips to provide the sailors with the necessary vitamin C.

You will use an iodometric titration method to determine the vitamin C content of a sample with a known amount of ascorbic acid, a store-bought vitamin C tablet, and an orange juice sample. The indicator system for this experiment will be an iodine/starch mixture. You will add a small amount of starch to the sample for which you want to determine the amount of vitamin C. Then you will start adding drops of iodine solution. As you begin, you will momentarily see a flash of blue color, which quickly disappears. In the presence of starch, iodine will turn a blueish color because of to the way in which these two molecules fit together. This color disappears quickly because the vitamin C will bind with the iodine and not permit it to react with the starch. When all of the vitamin C present in the sample is bound to the iodine, any additional iodine is free to react with the starch and create the bluish color. You know that you have achieved the reaction end point when the bluish color appears and remains.

There are three parts to this experiment. You will begin by standardizing the iodine indicator solution, because it is difficult to make an exact concentration. Then you

will determine the vitamin C content found in a purchased vitamin C tablet. Next you will determine the vitamin C content of a fruit juice sample. This lab does require some simple math. Be sure to keep good records and ask for help if you don't understand what is required.

Chemicals and Equipment

* You will need these items only if your laboratory instructor does not provide the solutions.

- 0.1 g ascorbic acid*
- 1.0 g starch*
- 0.6 g potassium iodide*
- 0.6 g iodine*
- 50 mL ethanol*
- 100-mL beaker*
- 150-mL beaker*
- 250-mL beaker*
- 600-mL beaker*
- 1-L plastic storage bottle*
- hot plate or Bunsen burner with ring stand, ring and wire gauze*
- glass stirring rod*
- standard ascorbic acid solution sample
- starch solution sample
- iodine solution sample
- vitamin C tablet
- fruit juice sample
- deionized or distilled water
- 100-mL volumetric flask with stopper or cap
- 24-well reaction plate—each well should be able to hold about 2.5 mL
- at least 2 graduated plastic pipets
- piece of white paper to place under the well plate to aid in color judgment
- balance
- weighing paper or weighing boat

Safety Concerns

- **Although none of the chemicals that you will be using are considered dangerous, use them carefully. Wash your hands and work surface if any are accidentally spilled.**
- **Goggles should be worn at all times.**

Procedure

Standardization of iodine solution

1. Obtain an ascorbic acid solution sample.
2. Using a clean graduated plastic pipet, repeatedly fill and empty the pipet with the ascorbic acid solution sample to rinse it thoroughly.
3. Using the rinsed pipet, add 1 mL of the ascorbic acid solution sample to three wells of a clean well plate.
4. Rinse the pipet thoroughly with water and use it to add one drop of starch solution to each well containing the ascorbic acid solution sample.
5. Obtain a sample of iodine solution.
6. To titrate each ascorbic acid solution sample, begin adding drops of iodine solution using a new pipet. This pipet must remain with the iodine solution throughout the experiment. Continue to add drops of iodine until you achieve a blue end point, where the blue color remains after a few seconds. Record the number of drops needed to reach the end point in the data table. Also record the color of your end point to the best of your descriptive ability. This should be repeated for all for three wells.

Determination of vitamin C content in a purchased tablet

7. To prepare the tablet, weigh the tablet and record its mass to the nearest milligram. Record the mass of the tablet in the data table. Also record the reported amount of vitamin C in the tablet (from the packaging of the tablets).
8. Add the tablet to a 100-mL volumetric flask. Fill the flask about halfway with water, and mix well until the tablet appears to be breaking apart and dissolving. Add sufficient water so that the bottom of the meniscus is at the etched line on the neck of the volumetric flask.

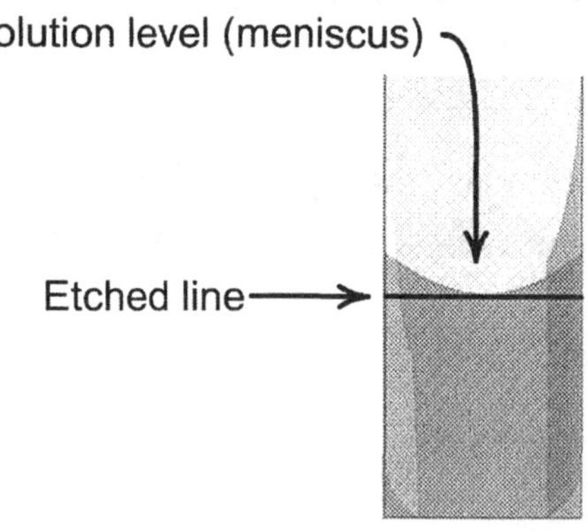

Figure 1. When viewed from the side, the bottom of the meniscus is level with the line etched in the volumetric flask.

9. Securely place the cap on the volumetric flask and repeatedly turn the flask upside down, mixing and swirling the solution to mix it thoroughly. Label the flask "vitamin C tablet solution." You may wish to do these three steps (steps 7–9) earlier in the experiment to allow the tablet to dissolve completely. Your lab instructor may have already done these steps prior to the lab session.
10. Using a clean graduated plastic pipet, repeatedly fill and empty the pipet with the vitamin C solution to rinse it thoroughly.
11. Using the rinsed pipet, add 1 mL of vitamin C solution to three wells of a clean well plate.
12. Rinse the pipet thoroughly with water and use it to add one drop of starch solution to each well containing the vitamin C solution.
13. To titrate each sample, begin adding drops of iodine solution, using the pipet from the standardization step. Continue to add drops until you achieve a blue end point. The blue may be slightly lighter than in the standardization step due to binders present in the vitamin C tablet; the binders are not water soluble and make the solution cloudy. Record the number of drops needed to reach the blue end point in the data table. Also record the color of your end point to the best of your descriptive ability. This should be repeated for all for three wells.

Determination of vitamin C content in a fruit juice sample

14. Obtain a sample of fruit juice. Record in your data table the identity of the juice and its reported vitamin C content.
15. Using a clean graduated plastic pipet, repeatedly fill and empty the pipet with the juice sample to rinse it thoroughly.

16. Using the rinsed pipet, add 1 mL of the juice sample to three wells of a clean well plate.
17. Rinse the pipet thoroughly with water and use it to add one drop of starch solution to each well containing the juice sample.
18. To titrate each sample, begin adding drops of iodine solution, using the pipet from the standardization step. Continue to add drops until you achieve a bluish end point. It is important to note that, due to the color of the juice sample, you will not achieve the same color end point as before. If you are using orange juice, the end-point color will likely be more of a green-blue color. You may need to do a few practice runs to determine the correct end-point color. Record the number of drops needed to reach the end point in the data table. Also record the color of your end point to the best of your descriptive ability. This should be repeated for all for three wells.

Stock solution directions

Follow these directions only if the standard ascorbic acid solution, the iodine solution and the starch solutions are not available from your laboratory instructor.

Standard ascorbic acid solution (0.1%)

- Weigh 0.1 g of ascorbic acid and add to a 150-mL beaker.
- Add 100 mL water and stir well.
- Label beaker "0.1% ascorbic acid solution."

Starch solution (1.0%)

- Weigh 1.0 g of starch and add to a 250-mL beaker with 100 mL of boiling water.
- Boil solution for one minute while stirring with a glass stirring rod.
- Remove from heat and allow the solution to cool, stirring occasionally. The solution will remain cloudy.
- Label beaker "1.0% starch solution."

Iodine solution

- Weigh 0.6 g potassium iodine and add to 500 mL of water in a 600-mL beaker. Stir well.
- Weigh 0.6 g iodine and add it to 50 mL of ethanol in a 100-mL beaker. Stir well.
- Add the iodine/ethanol solution to the potassium iodide solution in the 600-mL beaker. Stir well.
- Transfer this solution to a 1-L plastic storage bottle and add an additional 450 mL of water.
- Label the plastic storage container "iodine solution."

Cleanup and Disposal

- ☞ Be sure to wash thoroughly all of the equipment and supplies used during this lab.
- ☞ All excess solutions may be washed down the sink with generous amounts of water between each solution, or follow your instructor's directions.
- ☞ Be sure to wash your hands thoroughly at the end of this lab.

Calculations

Calibration factor

A calibration factor must be calculated for the actual iodine solution and equipment that you used in this experiment. This number may be different for each member of your lab class. The calculation is based on the data you collected in the first part of the experiment, in which the ascorbic acid solution contained 1 mg of vitamin C per 1 mL of solution.

$$\text{Calibration factor} = \frac{1\text{ mg ascorbic acid}}{1\text{ mL solution}} \times \frac{1\text{ mL of solution}}{X\text{ drops of iodine}}$$

The X represents the average number of drops of iodine that it took to reach the end point of the reaction.

Vitamin C tablet

$$\text{mg vitamin C in tablet} = \text{calibration factor} \times \frac{Y\text{ drops of iodine}}{1\text{ mL vitamin C tablet sol'n}} \times 100$$

The Y represents the average number of drops of iodine that it took to reach the end point of the reaction.

You multiply the value by 100 since you only used 1 mL out of the 100 mL of solution you prepared from the one tablet.

Fruit juice sample

$$\text{mg vitamin C in juice} = \text{calibration factor} \times \frac{Z\text{ drops of iodine}}{1\text{ mL juice sample}}$$

The Z represents the average number of drops of iodine that it took to reach the end point of the reaction.

Data and Observations

Standardization of the iodine solution

Trial #	1	2	3
Volume of ascorbic acid solution	1 mL	1 mL	1 mL
Drops of iodine			

End point color: ____________________________

Average number of drops of iodine required to reach end point: ____________________

Calibration factor: _____________ mg ascorbic acid / 1 drop iodine
(calculation instructions are after the data tables)

Determination of vitamin C content in a purchased tablet

Trial #	1	2	3
Volume of vitamin C tablet solution	1 mL	1 mL	1 mL
Drops of iodine			

Mass of vitamin C tablet: ____________________ grams

Reported vitamin C content in tablet: ________ milligrams or grams *(circle one)*

End-point color: ____________________________

Average number of drops of iodine required to reach end point: ____________________

Experimental vitamin C content in tablet: ________ milligrams
(calculation instructions are after the data tables)

Determination of vitamin C content in a fruit juice sample

Trial #	1	2	3
volume of fruit juice sample	1 mL	1 mL	1 mL
Drops of iodine			

Identity of fruit juice sample: ____________________

Reported vitamin C content in juice sample: ______ milligrams or grams *(circle one)*

Original fruit juice sample color: _________________

End-point color: _____________________________

Average number of drops of iodine required to reach end point: ____________________

Experimental vitamin C content in juice sample: ________ milligrams
(calculation instructions are after the data tables)

Post Experiment Questions

1. Grape juice, tomato juice and many other fruit juices contain vitamin C. Why were they not used in this experiment? How did you come to this answer?
2. Based on the procedures used in this experiment, do you believe that vitamin C is a fat-soluble vitamin or a water-soluble vitamin? How could you easily test your belief?
3. Why was it necessary to standardize the iodine indicator solution?
4. You recorded the amount of vitamin C reported in the juice sample and have now calculated the amount based on your experimentation. What is the percent difference between these two values?
5. You also recorded the amount of vitamin C in the tablet and have now calculated the amount based on your experimentation. What is the percent difference between these two values?

Fat Content Analysis of Chips and Nuts

Introduction

The purpose of this experiment is to determine the fat content in potato chips and peanuts. Fats will be extracted from the two samples using acetone. The mass of fat extracted from the food will be found after the acetone has evaporated.

Background Information

For many, "fat" is a taboo word. However, in terms of the normal functioning of the human body, fat is a necessary component. Fats are used to protect organs, store energy, and help absorb and move nutrients, and fats also play a part in the production of hormones. Fats play a role in the transport of cholesterol within the body. Cholesterol has been strongly linked to cardiovascular health. Dietary fats (fats that are consumed) are classified into two categories, saturated fats and unsaturated fats. Saturated fats are found in animal products, and the over consumption of these fats increases the risk of heart disease. Within the human body, saturated fat forms low-density lipoproteins (large molecules that are a combination of a fat molecule and a protein). Low-density lipoproteins (LDL) transport cholesterol around the body. Since cholesterol remains in the body when transported by LDL, high levels of LDL are considered a health risk. Unsaturated fats are found in vegetables and nuts, and they tend to have the beneficial effect of helping to clean arteries. The unsaturated fat forms high-density lipoproteins (HDL) that transport cholesterol to the liver, where it is prepared for excretion from the body. Since high-density lipoproteins transport cholesterol for disposal, higher levels of HDL are considered beneficial.

Both saturated and unsaturated fats are also known as **triglycerides** since they are esters of three fatty acid molecules and one glycerol molecule. The difference between saturated and unsaturated fats is in the types of fatty acids used to make the triglyceride.

While fats are essential for energy storage in the human body, excessive consumption of fatty foods leads to numerous health problems. Snack foods, like potato chips, chocolate, and nuts, all contain varying amounts of fats. In this experiment, the fat from peanuts and potato chips will be extracted and quantified. Fats are not soluble in water, but they are soluble in organic solvents like acetone. Once the fats are dissolved in acetone, they can be isolated by evaporating the acetone in a steam bath. By comparing the mass of fat extracted to the mass of food originally used, the fat content will be determined.

Chemicals and Equipment

- mortar and pestle
- two 50-mL beakers
- 10-mL graduated cylinder
- potato chips
- peanuts
- balance
- weighing paper
- steam bath, set up in a hood

Safety Concerns

- **Always wear safety goggles. You may also wish to wear gloves and an apron.**
- **Acetone is a flammable, organic solvent. Do not use any open flames when acetone is present. Work inside a hood when pouring acetone into a beaker.**

Procedure

1. Weigh the two clean 50-mL beakers that you will use to collect the fat from the peanuts and chips. Make sure the balance reads 0.000 g before placing the beakers on the balance. Record the mass to the nearest 0.001 g in the data table.
2. Using weighing paper, weigh a sample of potato chips to the nearest 0.001 g. Make sure you tare the balance after placing the weighing paper on the balance. You should use about 5 g of chips.
3. Place the chips in a mortar and use the pestle to crush the chips into small pieces.
4. Using a graduated cylinder, add 5 mL of acetone to the mortar. Mix the acetone with the chips using the pestle.
5. Carefully pour the acetone from the mortar into one of the preweighed 50-mL beakers. Record any color change in the acetone.
6. Add a second 5 mL of acetone to the mortar and mix thoroughly, again using the pestle. Carefully pour the acetone into the same beaker as in step 5.
7. Place the beaker with the acetone on a steam bath to evaporate the acetone. The steam bath should be set up in a hood.

8. The residue remaining in the beaker after the acetone has evaporated is the fat from the chips. Let the beaker cool and then weigh the beaker. Make sure the balance reads 0.000 g before placing the beakers on the balance. Record the mass to the nearest 0.001 g in the data table.

Cleanup and Disposal

- To clean the beaker, wash it thoroughly with warm soapy water.
- Repeat the procedure using peanuts instead of chips.
- The crushed chips and peanuts can be discarded in the trash.
- Any unused acetone should be discarded as directed by your instructor.
- All equipment should be washed thoroughly with warm soapy water.
- Be sure to wash your hands.

Data and Observations

Data table

Trial	1	2
Description of food		
Mass of beaker with fat (g)		
Mass of empty beaker (g)		
Mass of fat (g)		
Mass of food used (g)		
% of fat in food		

Post Experiment Questions

1. Which food, potato chips or peanuts, has the lower fat content?
2. Compare the fat content you found in the experiment to that listed on the package label of the foods you tested.
3. How did the appearance of the acetone change after you mixed it with the food?
4. This experiment is not designed to differentiate between saturated and unsaturated fats. Using the information that saturated fats make high-density products and unsaturated fats make low density products, how could you determine if the fat obtained is saturated or unsaturated fat?
5. Some snack foods are advertised to be lower in fats. Describe an experiment to test their claim and, if time permits, carry out your experiment.

Synthesis of Polymers

Introduction

Polymers are large organic molecules with masses that range from the thousands to millions of atomic mass units (amu). They consist of repeating identical molecules or units that we call monomers. You use many synthetic polymers every day. Plastic grocery bags are made of a high-density polyethylene (HPDE) film. The Teflon commonly found in cooking pans to make them nonstick is another everyday example of a polymer. Plastics and polyester, a component in many fabrics, are also synthetic examples. Polymers are also found in nature. For example, proteins, nucleic acids, and rubber are examples of natural polymers.

Background Information

One of the polymers you will synthesize in this laboratory exercise is nylon 66. This synthesis consists of the reaction of a di-acid chloride, a solution of adipoyl chloride in cyclohexane, with a diamine, a basic solution of aqueous hexamethylenediamine. The reaction of adipoyl chloride and the diamine produces the nylon polymer and hydrogen chloride, which turns into hydrochloric acid in aqueous solution. This is an example of a *condensation polymer.* The compound hydrogen chloride, or HCl, is produced in the process of the polymerization reaction. Many common polymers are condensation polymers that produce water, H_2O, instead of HCl. The production of water is the reason for the name condensation. Nylon 66 is only one example of nylon. In fact, nylon can be made from diamines that have longer $-CH_2-$ units in between the $-NH_2$ ends.

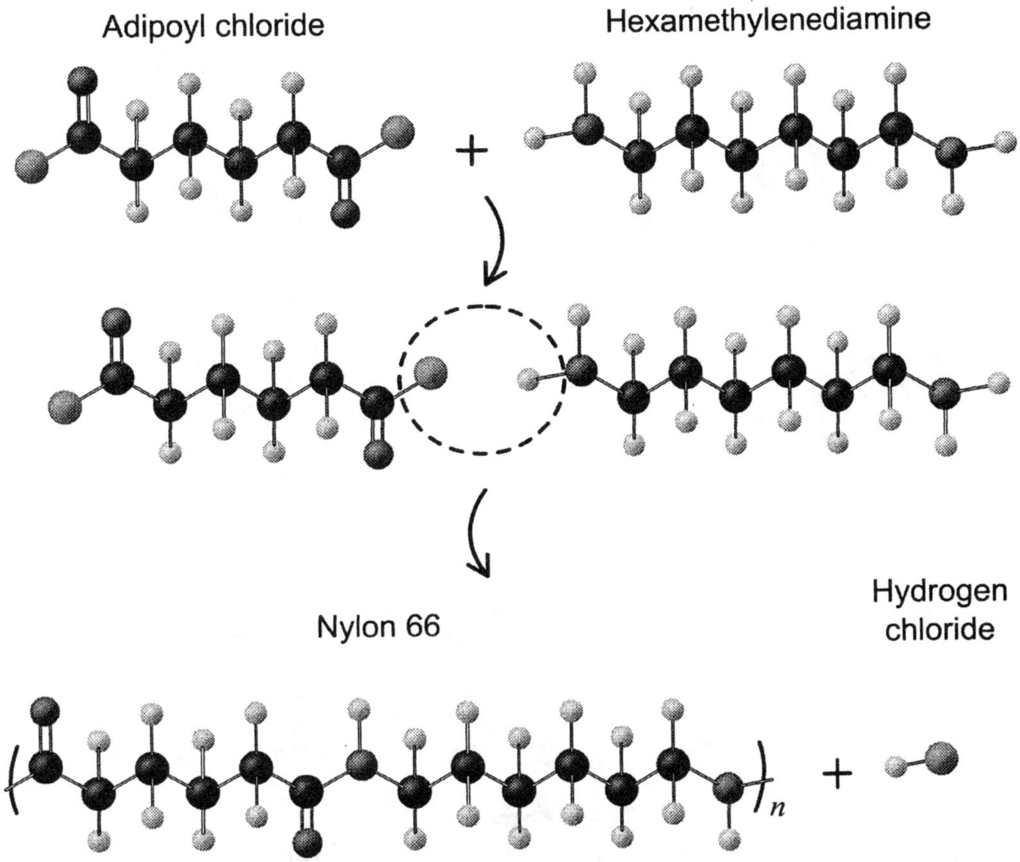

Figure 1. A ball-and-stick model showing the synthesis of nylon 66 from adipoyl chloride and hexamethylenediamine.

Nylon was first produced the in the mid 1930s by the DuPont Company in New Jersey. The first nylon product sold was a toothbrush that had bristles made from nylon strands. The most famous product made from nylon probably is the one that bears its name—nylons, or women's stockings—which first went on sale in May of 1940.

A second type of polymer you will synthesize is a borate cross-linked polymer made by reacting standard white Elmer's glue with a 4% sodium borate solution, $Na_2B_4O_7 \bullet 10H_2O$. Sodium borate is an everyday laundry product available at the grocery store as Borax. The final polymer that you will synthesize is made by reacting polyvinyl alcohol and sodium borate. By adding sodium borate to polyvinyl alcohol you are creating cross-links or bonds between adjacent polymer strands.

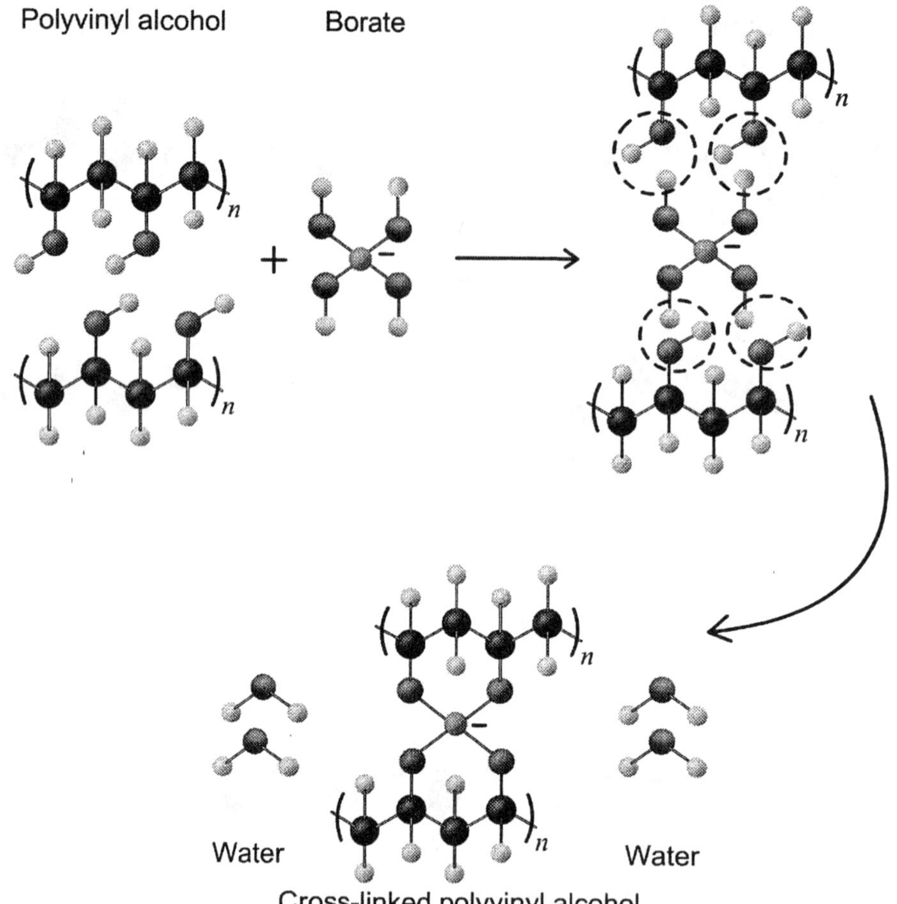

Figure 2. A ball-and-stick model showing a borate ion forming cross links between two polyvinyl alcohol polymers.

Creating more abundant and stronger bonds makes a more rigid and more viscous polymer cross-linking. You may recognize the product of this reaction as something that looks and behaves like silly putty! There are two different procedures to create the borate cross-linked polymers. After you have synthesized both of the products, you should be able to describe the differences between the two.

For each polymer you synthesize you will record several observations. Carefully examine the polymer sample and record its general appearance. Run your fingers over the sample's surface and record its texture. To test its flexibility, try bending the sample. To test its stretchability, pull the sample in two manners: first, pull the sample apart slowly; and second, pull the sample apart very quickly. Finally hold the sample about 12 inches above the lab bench, drop it, and see if it bounces. Record your results.

Chemicals and Equipment

- 25-mL or 50-mL graduated cylinder
- two 50-mL beakers
- 100-mL beaker
- 150-mL beaker
- glass stirring rod
- pair of tweezers
- small test tube or plastic dowel with ~0.5 cm diameter
- 10 mL adipoyl chloride/hexane solution—made by dissolving 4.6 grams of adipoyl chloride in 100 mL of hexane
- 10 mL basic aqueous hexamethylenediamine solution—made by dissolving 6 grams of hexamethylenediamine and 2 grams of sodium hydroxide in 100 mL water
- 25 mL Elmer's glue
- 10 to 20 mL aqueous 4% sodium borate solution
- 50 mL aqueous 4% polyvinyl alcohol solution

Safety Concerns

- **Always wear safety goggles. You may also wish to wear gloves and an apron.**
- **Avoid skin contact with the adipoyl chloride solution and the hexamethylenediamine solution. Adipoyl chloride/hexane solution is flammable and is toxic by ingestion and inhalation. The hexamethylenediamine is toxic by ingestion and is a strong tissue irritant. Should you get some on your skin, wash with copious amounts of water.**
- **Wash the nylon product well with water prior to handling. Although the washed nylon product should be safe, be sure to wash your hands when you are finished with this experiment.**

Procedure

Synthesis of nylon 66

1. Obtain 10 mL of the adipoyl chloride solution using your graduated cylinder and pour it into a 50-mL beaker that you have labeled #1.
2. Rinse your graduated cylinder well.
3. In a second 50-mL beaker, which you have labeled #2, add 10 mL of the hexamethylenediamine solution, which you measured with your graduated cylinder.
4. Very carefully and slowly, add the contents of beaker #2 into beaker #1 by gently pouring the contents down the side of beaker #1. *Do not stir or mix.* Two layers will form. The contents of beaker #1 will float on top of the contents of beaker #2. At the interface between the two layers, a nylon film will form.
5. Using tweezers grasp the nylon film that has formed at the interface and pull it straight up and out of the beaker. As you are pulling it out of the beaker, begin to wrap it around a test tube or plastic dowel. To continue pulling the nylon out of the beaker, begin to rotate the test tube/dowel. Continue pulling and rotating the tube/dowel until the original solutions are used up. If you are careful, you may be able to obtain one long continuous strand of nylon. However, if it does break, simply grasp the film between the two solutions and begin pulling it out again.
6. Do not handle the nylon until you have gently rinsed it with tap water. After you have gently rinsed your nylon strand with tap water, record your observations and let it sit out to dry.
7. Once the nylon has dried, record your observations.
8. The dried nylon strands may be disposed of in the trash. Any excess original solutions should be placed in the waste containers as directed by your instructor. Do not forget to wash thoroughly all glassware and tools.

Synthesis of a borate polymer

9. Using your graduated cylinder, measure 25 mL of Elmer's glue and add it to a 150-mL beaker. Wash your graduated cylinder.
10. Measure 20 mL of water and add it to the Elmer's glue, stirring well.
11. While stirring continually with a stirring rod, add aqueous 4% sodium borate solution until no more is absorbed. At this point the polymer will develop into a

firm yet flexible ball. You should need between 5 mL and 10 mL. Be careful not to add too much.

12. Remove your polymer ball and allow it to sit on the lab bench for a few minutes. It will be sticky at first.
13. Record your observations.
14. The polymer ball can be disposed of in the trash can. Any excess solution can be disposed of in the sink with water. Be sure to wash thoroughly all glassware and tools.

Synthesis of a polyvinyl alcohol gel

15. Using your graduated cylinder, obtain 50 mL of aqueous 4% polyvinyl alcohol solution and place it in a 100-mL beaker. Wash your graduated cylinder.
16. Using your graduated cylinder, obtain 5 mL of aqueous 4% sodium borate solution and, while continuously stirring, add it to the beaker. This may require several minutes of stirring.
17. Once the two ingredients are fully incorporated, remove the polymer from the beaker and record your observations. Be sure to wash your hands after handling the sample.
18. The polyvinyl alcohol gel may be disposed of in the trash. Any remaining reactants should be placed in a waste container as directed by your instructor. Be sure to wash thoroughly all glassware and tools.

Cleanup and Disposal

- The nylon strands, glue-borate polymer, and polyvinyl alcohol gel can be placed in the trash can. Any remaining solutions should be placed in the waste containers or down the sink as directed by your instructor.
- Be sure to wash thoroughly all glassware and tools.
- Be sure to wash your hands thoroughly.

Data and Observations

Nylon synthesis

Data table		
Test	Wet nylon	Dry nylon
General appearance		
Texture		
Flexibility		
Stretchability		
How well does it bounce?		

Borate cross-linked polymer

Data table

Test	**Borate cross-linked polymer**
General appearance	
Texture	
Flexibility	
Stretchability	
How well does it bounce?	

Polyvinyl alcohol gel

Data table

Test	**Polyvinyl alcohol gel**
General appearance	
Texture	
Flexibility	
Stretchability	
How well does it bounce?	

Post Experiment Questions

1. There are different types of nylon that can be synthesized. Why is the one that you synthesized referred to as "nylon 66"?
2. Nylon is a common component in many modern fabrics. Do your observations coincide with its stretchable and flexible properties when found in fabrics?
3. In the chemical reaction for the synthesis of nylon, HCl is shown as a product. Did you observe the production of HCl? Since HCl is a strong acid, did you take any special precautions?
4. The final two polymers you made are commonly called homemade silly putty. One fun property of silly putty is its ability to pick up an image from a newspaper that uses water-soluble ink. Do your homemade polymers possess this property? If you do not have a newspaper, how can you test this?
5. How did the general appearance of the polyvinyl alcohol change after it was polymerized with the sodium borate solution?

The Isolation of DNA

Introduction

The purpose of this experiment is to extract DNA from two sources—an onion and a kiwi fruit—and compare the DNA from the two sources. For both DNA sources, you will mechanically break the cell walls and continue to break them down by adding a homogenizing agent. You will precipitate the DNA by adding isopropyl alcohol, and finally you will separate the DNA, which will have the appearance of stringy egg whites. Throughout the experiment, you should carefully record your observations.

Background Information

Deoxyribonucleic acid, or DNA, is a complex molecule that everyone has heard of. You hear about it on the daily news and in popular entertainment programs because it relates to crime scene investigations. It may also be discussed with such hot-button topics as genetically modified foods, genetic engineering, and the potential to cure many diseases. Serving as the blueprint for life, DNA is found in the nucleus of all living cells.

DNA is a complex molecule that is actually a polymer consisting of thymine, adenine, guanine, cytosine, phosphate, and deoxyribose. The phosphate and deoxyribose fragments help to construct the backbone of the DNA. The four nucleotide bases—thymine, adenine, guanine, and cytosine—are attached to the backbone, and their specific order is the genetic code for the organism. The DNA forms a double helix; that is two strands of DNA are bound together and twisted in a helical pattern. To form the double helical structure of the DNA polymer, different pairs of nitrogenous bases from each chain are attracted to each other via hydrogen bonding. Specifically, adenine and thymine are attracted to each other, and guanine and cytosine are attracted to each other.

Cells are protected from their surrounding environment by a cell wall. Within a cell, the nucleus is separated from the rest of the cell by a nuclear membrane. In order to isolate DNA, both the cell wall and nuclear membrane must be broken down. Liquid laundry detergent will break down the cell walls of the onion and kiwi, thereby releasing the DNA from inside the cells. DNA molecules are fragile and are easily broken apart when they are removed from the cell. A solution, called a **homogenizing solution**, is made to keep the DNA from breaking apart. The homogenizing solution is salty and slightly basic—an environment close to that of living cells—and therefore the DNA will easily dissolve. The homogenizing solution is cooled to help stabilize the system. A well-cooled homogenizing solution will slow down the rate of DNA breakup.

Chemicals and Equipment

- deionized or distilled water
- crushed ice
- 1.5 g sodium chloride, NaCl
- 5.0 g sodium bicarbonate, $NaHCO_3$
- 5.0 mL liquid laundry detergent
- 100 mL isopropyl alcohol
- small onion or half a large onion
- 1 kiwi fruit
- balance (accurate to 0.1 g)
- weighing paper or weighing boat
- two 100-mL beakers
- two 250-mL beakers
- two 1000-mL beakers
- 25-mL or 50-mL graduated cylinder
- glass stirring rod
- filter paper
- funnel
- ring stand and ring
- mortar and pestle
- knife or chopping device

Safety Concerns

- **Although none of the chemicals that you will be using are considered dangerous, use them carefully. Wash your hands and work surface if anything is accidentally spilled.**
- **Goggles should be worn at all times.**

Procedure

Preparation of a homogenizing solution

1. Using a balance and weighing paper, weigh 1.5 g of sodium chloride, NaCl, and transfer it to a 250-mL beaker.
2. Using a balance and weighing paper, weigh 5.0 g of sodium bicarbonate, $NaHCO_3$, and add it to the beaker containing the sodium chloride.
3. Using a graduated cylinder, measure 120 mL deionized water and add it to the beaker.
4. Using a graduated cylinder, measure 5.0 mL of liquid laundry detergent and add it to the beaker.
5. Stir this solution well using a stirring rod.
6. Prepare an ice bath by adding crushed ice to a large (1000-mL) beaker with some water. Be careful not to add too much water. If you have more water than ice, the beaker with the homogenizing solution will sink, water will flow into the beaker, and you will need to start over.
7. Carefully place the 250 mL beaker containing homogenizing solution into the ice bath. Set the ice bath containing the homogenizing solution aside. The homogenizing solution needs to be very close to $0^{o}C$ when you add it to the onion or kiwi pulp.

Isopropyl alcohol preparation

8. Prepare a second ice bath using a 1000-mL beaker, crushed ice and a small amount of water.
9. Using a graduated cylinder, measure 100 mL of isopropyl alcohol and add it to a 250-mL beaker. Carefully place the beaker in the ice bath. Be careful not to allow this beaker to sink in the ice bath.
10. Set the ice bath containing the isopropyl alcohol aside. The isopropyl alcohol temperature needs to be very close to 0 oC when the isopropyl alcohol is added to the DNA containing liquid.

DNA extraction from onion

11. Obtain a small onion or half of a large onion. Dice the onion using a knife.
12. Use a mortar and pestle to mash the pieces of onion into a pulp. This will begin to break up the cell walls.
13. Transfer approximately 20.0 mL of the onion pulp to a 100-mL beaker and add 40.0 mL of the well-cooled homogenizing solution.
14. Stir this mixture vigorously with a stirring rod for about 4 minutes. The detergent in the homogenizing solution will continue to break down the cell walls, thereby releasing the DNA from inside the cells.
15. Place this beaker in one of your already prepared ice baths for 5 minutes. The solids will settle to the bottom of the beaker. The DNA is contained in the liquid at the top.
16. Set up a funnel in a ring on a ring stand. Fold a piece of filter paper and place it in the funnel. Place a 100-mL beaker under the funnel to collect the filtrate. Label the beaker "filtrate."
17. Carefully decant the liquid into the funnel. Try to keep as much of the solid material in the beaker as possible. The filter paper will filter out any solids but allow the DNA molecules and fragments to pass through. Rinse the solids remaining in the beaker with an additional 5 mL of the cold homogenizing solution. Decant the solution into the funnel. The filter paper and solid materials should be thrown away in the trash can. Do not put any of the solid materials in the sink.
18. Add enough well-cooled isopropyl alcohol to the DNA-containing filtrate so that you have approximately 2 inches of isopropyl alcohol sitting above the DNA-containing filtrate. As you add the well-cooled isopropyl alcohol, you want to maintain two distinct layers. You want as little mixing between the two liquids as possible. By letting the isopropyl alcohol slowly run down the side of the beaker, you will decrease the mixing between the two liquids. The DNA molecules and fragments will be attracted to the interface between the two liquids.
19. Carefully insert a clean glass stirring rod through the isopropyl alcohol layer down into the DNA-containing liquid. Carefully begin to rotate the glass stirring rod in one direction. The polar ends of the DNA molecules will be attracted to the polar glass stirring rod. As you rotate the glass stirring rod, the DNA molecules should begin to wind onto it. Continue winding the DNA onto the glass stirring rod for about 1 minute.
20. Record your observations for this entire experimental section in the data table.

DNA extraction from kiwi

Repeat steps 11–20 using a kiwi in place of an onion. Record your observations for this entire experimental section in the data table.

Cleanup and Disposal

- Be sure to wash thoroughly all of the equipment and supplies used during this lab.
- All excess solutions may be washed down the sink with generous amounts of water or follow your instructors directions.
- Excess onion or kiwi fruit pulp should be thrown in the trash, not put in the sink.
- Be sure to wash your hands thoroughly at the end of this lab.

Data and Observations

Be sure to record observations before, during, and after each laboratory experiment section.

Observations during DNA extraction from onion

Appearance of DNA extracted from onion

Observations during DNA extraction from kiwi fruit

Appearance of DNA extracted from kiwi

Post Experiment Questions

1. Based on the experimental procedure and your observations, do you believe that DNA is water soluble or soluble in isopropyl alcohol? What led you to this decision?
2. What similarities and what differenced did you observe between the onion and kiwi DNA?
3. How does the color of the plant correspond to the color of the DNA?
4. Using your textbook as a resource, draw the four nucleic acid molecules present in DNA. As a reminder, they are guanine, cytosine, adenine, and thymine. What do these four molecules have in common?
5. You extracted DNA from two plant sources. Would you expect to follow the same procedure and obtain the same results from any living cell? What led you to this decision?

Catalytic Breakdown of Hydrogen Peroxide

Introduction

The purpose of this experiment is to investigate the activity of the enzyme catalase under different reaction conditions. The activity of the enzyme will be measured by collecting the gas produced by the catalase-mediated decomposition of hydrogen peroxide.

Background Information

Thousands of chemical reactions occur within living organisms. Specialized proteins called **enzymes** catalyze almost all of these reactions. Enzymes are large molecules that have a specific structure that allows specific reactants to come together in the proper orientation to react. Living organisms are dependent upon the proper functioning of enzymes to survive.

One prevalent enzyme found in plants, animals, and bacteria is catalase. Catalase catalyzes the reaction of two hydrogen peroxide molecules to produce water and oxygen.

$$H_2O_2 + H_2O_2 \xrightarrow{\text{catalase}} H_2O + O_2$$

Hydrogen peroxide is a very reactive compound that can damage cells. It is produced in the process of metabolism and is used by white blood cells to destroy foreign bacteria in the body. The catalase enzyme breaks down hydrogen peroxide within a cell before the hydrogen peroxide can cause any damage.

The unique structure of enzymes allows them to be specific as to which reactions they catalyze. However, if the structure of the enzyme is disturbed, then the functionality of the enzyme is lost, and the enzyme no longer catalyzes the reaction. Enzymes are proteins, and like other proteins, their structure is affected by the temperature, acidity (or pH), and ionic strength (or salt concentration) of the solution. Increasing the temperature of an enzyme solution breaks the weak bonds that give the enzyme its unique structure. Once these bonds are broken, the enzymes, like other proteins, lose their conformation and tend to coagulate into a solid mass. The hydrogen ion concentration, or pH, of the solution can have similar effect. Excess hydrogen ions interfere with the internal bonding of the enzymes, causing the enzymes to lose their conformation.

Although enzymes are specific as to the reactions they can catalyze, they are not perfect. Enzyme activity can be inhibited by other compounds. Inhibition of enzyme activity occurs in one of two processes, competitive inhibition and noncompetitive inhibition. Noncompetitive inhibition occurs when other compounds can bind to the enzyme, causing the enzyme to lose its functionality completely. Competitive inhibition occurs when other compounds can fit into the enzyme, substituting for the normal reactant. In competitive inhibition, the enzyme continues to function, but the normal

reaction is slowed because the reactants have to compete with the inhibiting compounds for space in the enzyme.

In this experiment, the catalase enzyme extracted from potatoes will be used to catalyze the breakdown of hydrogen peroxide into water and oxygen gas. The affects of temperature on enzyme activity will be investigated. To measure the activity of the enzyme, the volume of oxygen gas will be collected. Collecting gas released in a chemical reaction is a common procedure and can be effectively accomplished by inverting a graduated cylinder of water in a larger vessel of water and channeling the gaseous product into the graduated cylinder through rubber tubing. The gas will displace water from the graduated cylinder, which allows the volume of gas produced to be determined.

Chemicals and Equipment

- 1000-mL beaker
- four 250-mL beakers
- 600-mL beaker
- 100-mL graduated cylinder
- 10-mL graduated cylinder
- 8″ test tube
- rubber stopper with glass tube (fits large test tube)
- Rubber tubing
- Ring stand
- ring and mesh wire
- clamp
- knife (to cut potato)
- potato
- 3% hydrogen peroxide solution
- hot plates

Safety Concerns

- **Always wear safety goggles.**
- **Hydrogen peroxide is a strong oxidant and will cause vomiting if ingested.**
- **Wash your hands before leaving the lab.**

Procedure

1. Prepare the gas collection assembly shown in Figure 1.

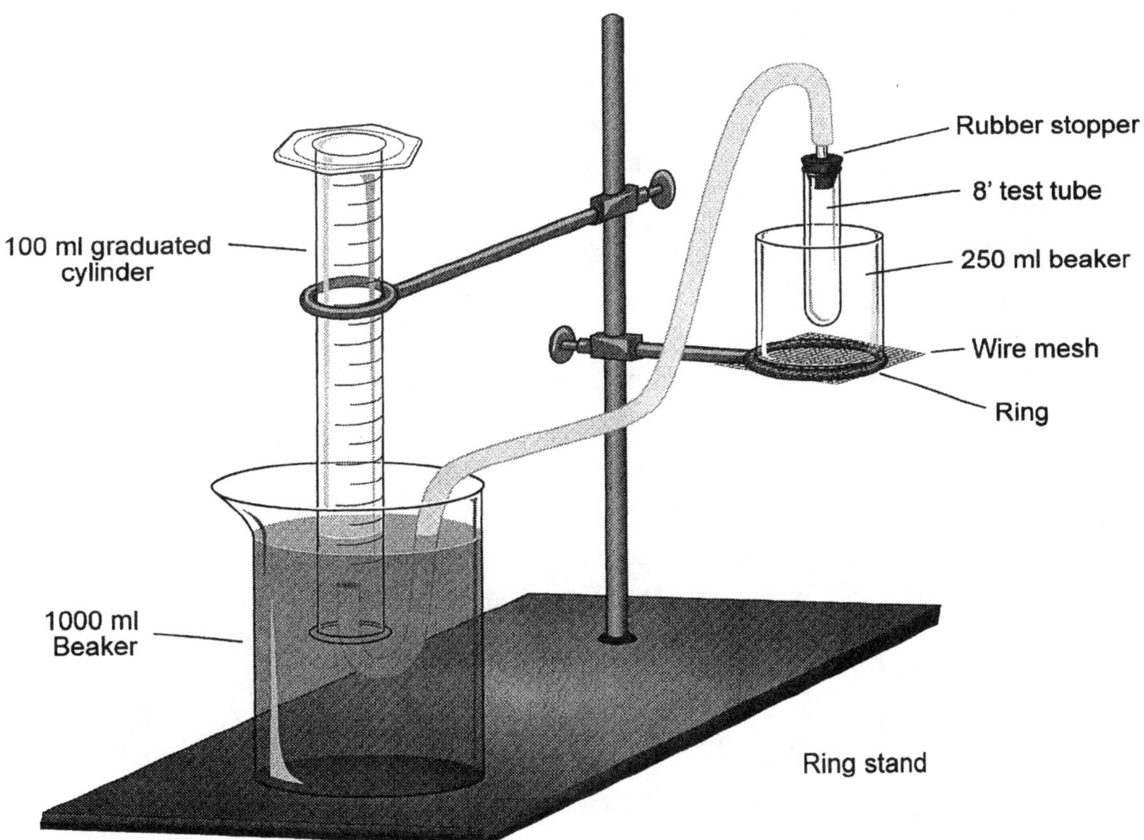

Figure 1. Gas collection assembly

Once the tubing is placed in the 1000 mL beaker, make sure the end connected to the rubber stopper remains above the water level in the beaker. Also, keep in mind that water will be displaced from the graduated cylinder during the reaction; make sure there is enough volume in the beaker to accommodate more water!

2. Prepare a hot water bath by placing a 250-mL beaker with 200-mL of water on a hot plate. The temperature of the water bath should be kept between 35°C and 37°C.
3. Prepare a second hot water bath by placing a 250-mL beaker with 200-mL of water on a hot plate. The temperature of the water bath should be kept between 65°C and 70°C.
4. Prepare a cold water bath by mixing ice and water in a 250-mL beaker. To maintain a temperature below 5°C, place the beaker in a larger vessel filled with ice.
5. Slice a raw potato into 5-mm cubes.
6. Fill a third of a large test tube with potato pieces. Make a line on the test tube to mark the height of the potato pieces for easy reference in later trials.
7. Measure 5-mL of hydrogen peroxide in a graduated cylinder and pour it into a small test tube.
8. Read the volume of water remaining in the 100-mL graduated cylinder. Record the volume in the data table.
9. Add the 5-mL of hydrogen peroxide to the potato pieces and place the rubber stopper on the test tube. Start timing the reaction. Record your observations.
10. After the reaction slows, record the amount of gas collected in the inverted graduated cylinder. Also record the time that has elapsed.
11. Disconnect the rubber stopper from the test tube; remember to keep the end of the tube above water level. Dispose of the potato pieces as directed by your instructor.
12. Repeat steps 5–11 with new potato pieces. In this trial, place the 5-mL of hydrogen peroxide in the 35°C hot water bath for 3 minutes prior to adding it to the potato. While the solution is reacting, take the beaker from the hot plate and place it on the ring stand. Place the test tube in the 35°C water-bath beaker while the reaction occurs. Depending on the level of water remaining in the inverted graduated cylinder, you may have to refill the graduated cylinder with water.
13. Repeat steps 5–11 with new potato pieces. In this trial, place the 5-mL of hydrogen peroxide in the 65°C hot water bath for 3 minutes prior to adding it to the potato. While the solution is reacting, take the beaker from the hot plate and place it on the ring stand. Place the test tube in the 65°C water-bath beaker while the reaction occurs. Depending on the level of water remaining in the inverted graduated cylinder, you may have to refill the graduated cylinder with water.
14. Repeat steps 5–11 with new potato pieces. In this trial, place the 5-mL of hydrogen peroxide in the cold water bath for 3 minutes prior to adding it to the potato. While the solution is reacting, take the beaker from the cold water bath

and place it on the ring stand. Place the test tube in the cold-water bath beaker while the reaction occurs. Depending on the level of water remaining in the inverted graduated cylinder, you may have to refill the graduated cylinder with water.

Cleanup and Disposal

- All equipment should be washed thoroughly with warm soapy water.
- Excess hydrogen peroxide should be discarded as directed by your instructor.
- Potato pieces can be discarded in the trash unless otherwise specified by your instructor.

Data and Observations

Data table

Trial	Room temperature	Hot water bath 1	Hot water bath 2	Cold water bath
Temperature of reaction (°C)				
Time elapsed (sec)				
Final volume of water (mL)				
Initial volume of water (mL)				
Volume of gas collected (mL)				

Post Experiment Questions

1. What effect did the temperature of the peroxide solution have on the catalase activity?
2. What is the purpose of the two hot water baths?
3. Mild hypothermia occurs when a person's body temperature drops just a couple of degrees Fahrenheit. Severe hypothermia occurs when a person's body temperature drops 8°F to 12°F, causing the body to shut down into a metabolic coma. Based upon your observations in this lab, how does enzyme activity change over this relatively small temperature range?
4. Copper sulfate reacts with the active site in the catalase enzyme, causing the enzyme to change. What type of inhibition does copper sulfate cause, and how would this inhibition affect the oxygen produced in the experiment?
5. Will the enzyme activity return to normal after the potato has been heated to 65°C? Will the enzyme activity return to normal after the potato has been cooled to 5°C? Design an experiment to test these situations and, if time permits, carry out the experiment.

NATURAL INDICATORS: DETERMINING PH USING PRODUCE

Introduction

The goal of this experiment is use a pH indicator made from purple cabbage to determine the pH of different household substances. This experiment is divided into three parts. In part 1 you will make your purple-cabbage indicator solution. In part 2 you will make and test seven solutions of known pH. This will establish color standards for the seven pH values. Finally, in part 3 you will determine the pH of seven common household substances using the purple cabbage juice as the indicator.

Background Information

You may have heard of the term **pH** long before your started studying chemistry. You may have seen advertisements for deodorant that is pH balanced for a woman. You may have shampoos that are advertised as being pH balanced. The pH of a substance is the measure of its acidity. The acidic and basic characteristics of many substances influence their purpose and use in our everyday lives.

Recall from lecture discussions that an acid is a substance that produces H^+ in aqueous solutions and a base is a substance that produces OH^- in aqueous solutions. pH is a measure of the concentration of H^+ present in a solution. The more acidic a solution is, the higher the H^+ concentration. This results in a lower pH value. The more basic a solution is, the lower the H^+ concentration. This results in a higher pH value. Acids generally have pH values that range from 1 to 7. Stronger acids have lower pH values. Weaker acids have pH values closer to 7. Bases generally have pH values that range from 14 to 7. Stronger bases have higher pH values while weaker bases are closer to 7. A pH of 7 indicates that the solution is neutral.

An indicator is a chemical substance that changes color based on the pH of the substance being tested. A common indicator is litmus. This chemical substance is blue when in a basic solution and red when in an acidic solution. Litmus does not display the range of colors that are necessary for today's experiment. You will make and use purple cabbage juice as your indicator.

Purple cabbage contains a class of chemicals called anthocyanins. When mixed with an acidic sample, purple cabbage juice will change to a pink or red color. When mixed with a neutral sample, it remains purple. When mixed with a basic sample, it will become blue or green.

Chemicals and Equipment

- 10 mL of 0.1 M HCl
- 10 mL of 0.1 M NaOH
- distilled or deionized water
- approximately a quarter of a head of purple cabbage
- 10 mL of vinegar
- 10 mL of lemon juice
- 10 mL of noncolored clear soft drink
- 10 mL of common cleaning ammonia
- 10 mL of noncolored clear shampoo
- 10 mL of noncolored clear liquid hand soap
- 10 mL of liquid drain opener
- 10-mL graduated cylinder
- 25-mL graduated cylinder
- 100-mL graduated cylinder
- graduated or volumetric beryl-style pipet
- two 500-mL beakers
- 7 100-mL beakers
- 14 test tubes
- test tube rack
- pair beaker tongs
- hot plate or Bunsen burner set up (Bunsen burner, ring stand, ring support, and wire gauze pad)

Safety Concerns

- **You will be using several strong acids, strong bases, weak acids, and weak bases. If by accident you get any on your skin, wash with copious amounts of water and notify your instructor immediately. If you accidentally spill any chemical on your workspace, clean it up immediately.**
- **Goggles should be worn at all times.**

Procedure

Part 1: Making your indicator solution

You will make a natural indicator solution from common purple cabbage.

1. Obtain your cabbage sample and tear or cut it into small pieces, each piece about the size of a quarter. You should use approximately one quarter of a head of a small purple cabbage.
2. Add the cabbage pieces and approximately 250 mL of deionized water to a 500-mL beaker.
3. Using a Bunsen burner setup or a hot plate, bring your cabbage/water mixture to a boil. Allow the mixture to boil until the water turns a dark purple color.
4. Remove your hot beaker from the Bunsen burner or the hot plate and allow it to cool completely.
5. Decant the indicator solution by pouring off the liquid into a clean 500-mL beaker, leaving the cabbage solids behind. The remaining cabbage solids can be disposed of in the trash can.

Part 2: Determining standard color changes

To help you determine the pH of various samples, you need to determine what color changes occur at several pH values. You will obtain stock solutions of hydrochloric acid, HCl, and sodium hydroxide, NaOH, and dilute them several times to create solutions that will have several pH values.

6. Using the 25-mL graduated cylinder, obtain 10 mL of 0.1 M HCl solution. Place it in a 100-mL beaker labeled pH=1. Wash the graduated cylinder thoroughly.

7. Using the graduated beryl-style pipet, remove 1.0 mL of the 0.1 M HCl solution and place it in a 100-mL graduated cylinder. Wash the pipet thoroughly.
8. To this graduated cylinder, add enough deionized water to make the total volume 100 mL. Pour this solution into a 100-mL beaker labeled pH=3. Wash the graduated cylinder thoroughly.
9. Using the pipet, remove 1.0 mL of the pH=3 solution and transfer it to the 100-mL graduated cylinder. Wash the pipet thoroughly.
10. Once again, add enough deionized water to make the total volume 100 mL.
11. Transfer this solution to a 100-mL beaker labeled pH=5. Wash the graduated cylinder thoroughly.
12. Using a Bunsen burner setup or a hot plate, boil approximately 50 mL of deionized water in a 100-mL beaker and allow it to cool. Once it has cooled completely, label the beaker as pH=7.
13. Using the 25-mL graduated cylinder, obtain 10 mL of 0.1 M NaOH solution. Place it in a 100-mL beaker labeled pH=13. Wash the graduated cylinder thoroughly.
14. Using the graduated beryl-style pipet, remove 1.0 mL of the 0.1 M NaOH solution and place it in a 100 mL graduated cylinder. Wash the pipet thoroughly.
15. To this graduated cylinder, add enough deionized water to make the total volume is 100 mL. Pour this solution into a 100-mL beaker labeled pH=11. Wash the graduated cylinder thoroughly.
16. Using the pipet, remove 1.0 mL of the pH=11 solution and transfer it to the 100-mL graduated cylinder. Wash the pipet thoroughly.
17. Once again, add sufficient deionized water to make the total volume is 100 mL.
18. Transfer this solution to a 100-mL beaker labeled pH=9. Wash the graduated cylinder thoroughly.
19. Label seven small test tubes with pH=1, pH=3, pH=5, pH=7, pH=9, pH=11, and pH=13. Place them in order in a test tube rack.
20. To the test tube labeled pH=1, add approximately 5.0 mL of the pH=1 solution using a 10-mL graduated cylinder; add approximately 5.0 mL of the cabbage-juice indicator using the same graduated cylinder. Record the resulting color in the data table. Be sure to wash the graduated cylinder thoroughly each time you use it.
21. To the test tube labeled pH=3, add approximately 5.0 mL of the pH=3 solution using a 10-mL graduated cylinder; add approximately 5.0 mL of the cabbage-juice indicator using the same graduated cylinder. Record the resulting color in

the data table. Be sure to wash the graduated cylinder thoroughly each time you use it.

22. To the test tube labeled pH=5, add approximately 5.0 mL of the pH=5 solution using a 10-mL graduated cylinder; add approximately 5.0 mL of the cabbage-juice indicator using the same graduated cylinder. Record the resulting color in the data table. Be sure to wash the graduated cylinder thoroughly each time you use it.
23. To the test tube labeled pH=7, add approximately 5.0 mL of the pH=7 solution using a 10-mL graduated cylinder; add approximately 5.0 mL of the cabbage-juice indicator using the same graduated cylinder. Record the resulting color in the data table. Be sure to wash the graduated cylinder thoroughly each time you use it.
24. To the test tube labeled pH=9, add approximately 5.0 mL of the pH=9 solution using a 10-mL graduated cylinder; add approximately 5.0 mL of the cabbage-juice indicator using the same graduated cylinder. Record the resulting color in the data table. Be sure to wash the graduated cylinder thoroughly each time you use it.
25. To the test tube labeled pH=11, add approximately 5.0 mL of the pH=11 solution using a 10-mL graduated cylinder; add approximately 5.0 mL of the cabbage-juice indicator using the same graduated cylinder. Record the resulting color in the data table. Be sure to wash the graduated cylinder thoroughly each time you use it.
26. To the test tube labeled pH=13, add approximately 5.0 mL of the pH=13 solution using a 10-mL graduated cylinder; add approximately 5.0 mL of the cabbage-juice indicator using the same graduated cylinder. Record the resulting color in the data table. Be sure to wash the graduated cylinder thoroughly each time you use it.

Keep these standards so that they can be used to determine the pH of various substances.

Part 3: pH determination of various substances

You will now determine the pH of various common household substances. You will test vinegar, a noncolored clear soft drink, lemon juice, common cleaning ammonia, a noncolored clear shampoo, a noncolored clear liquid hand soap, and liquid drain opener.

27. For each sample, use the 10-mL graduated cylinder to obtain and transfer 5.0 mL of the sample to a test tube and also to add 5.0 mL of cabbage juice indicator. Wash the graduated cylinder thoroughly.
28. Record the resulting color in the data table.
29. Compare this color to your standard colors and pH values. Determine the pH of the sample and record it in the data table.

Cleanup and Disposal

- Be sure to wash thoroughly all of the equipment and supplies used during this experiment.
- Leftover cabbage and cabbage solids should be disposed of in the trash can.
- All solutions can be disposed of in the sink; add generous amounts of water to flush down each sample before you add the next sample, unless otherwise directed by your instructor.
- Be sure to wash your hands thoroughly at the end of this experiment.

Data and Observations

Standard pH Solutions

Data table

Solution pH	Color
1	
3	
5	
7	
9	
11	
13	

Sample Solutions

Data table

Substance	Color	pH
Vinegar		
Lemon juice		
Soft drink		
Cleaning ammonia		
Shampoo		
Liquid hand soap		
Liquid drain opener		

Post Experiment Questions

1. Consider the pH of your shampoo sample. What do you think it means when a shampoo claims that it is pH balanced for your hair? Why do you think so?
2. Consider the pH of your soft drink and the fact that your stomach contains hydrochloric acid, HCl, which is a strong acid. If you consume a large amount of soft drink in a short period of time, you are likely to get an upset stomach. To solve this problem you might take an antacid tablet. What do you believe the approximate pH range of the antacid tablet is? What type of reaction would occur in your stomach when the antacid tablet is added to your overly acidic stomach? Why do you think so?
3. Flower color can be influenced by pH. Most hydrangea flowers are blue or purple when they are in acidic soil and pink when they are in a basic soil. Do a little Internet research. What chemical substance in hydrangea flowers causes this reaction and color change?
4. Based on your results in the experiment and the purpose of the samples you tested, what do you think the pH range is of household cleaners, detergents, soaps, and shampoos? Why do you think so?
5. Frequent hand washing can dry out your skin. This is due to the pH of the soap. Based on this, do you think that your skin is acidic or basic? Why do you think so?

solutions are acids and which are bases, what their concentrations are, and in which solutions the indicator is located. Remember that the indicator is colorless in the presence of an acid and pink in the presence of a base. When you are mixing solutions, carefully watch for color changes. If a faint pink color develops and then quickly disappears, you have not reached the point of neutralization. Neutralization is reached when the pink color remains for about 15 seconds. As you mix the unknowns, only mix two at a time and be sure to count the drops you add. Counting the drops will help you determine the concentrations.

For example, if you begin by adding one drop of solution A to a well and then begin adding drops of solution B and the resulting solution changes to pink and remains pink after one drop of B, you know that you have determined that the indicator is present in solution A. You also now know that A is an acid and B is a base. Additionally, you have determined that A and B have the same concentration, since it was a one-to-one ratio of drops of A to drops of B. You will need further experimentation to determine if their concentrations are both 0.1 M, 0.2 M, 0.4 M, or 0.8 M.

Another example would have you begin by again adding one drop of A to a well. Now you start adding drops of C. After you have added one drop of C, a faint pink color develops but quickly disappears, so you add a second drop. When the second drop is added the pink color remains. You know that C is a base and is half the concentration of A. To determine their exact concentrations, further experimentation is needed.

If no reaction occurs (because no pink color develops and remains) you will know that the two solutions are either both acidic or both basic.

The process of determining which solutions are acids, which are bases, which contain the indicator, and what all of the concentrations are is time-consuming. You will need to keep your work area organized and keep accurate records of your procedure and your results.

Cleanup and Disposal

- ☞ Be sure to wash thoroughly all of the equipment and supplies used during this lab.
- ☞ Your instructor will provide you with directions for either washing or disposing of the beryl-style pipet.
- ☞ Be sure to wash your hands thoroughly at the end of this lab.

Data and Observations

To report your results, you will be asked either to develop an appropriate data table or to provide a detailed procedure with the results. For this reason we have not provided you with a data table. Be sure to ask your instructor before beginning the experiment how he or she wants your results to be presented.

Post Experiment Questions

1. Which solutions contained the phenolphthalein indicator?
2. Was the indicator in an acidic sample or in a basic sample? Why was it placed in that species and not in the other?
3. Identify the eight solutions as one of the following:

0.1 M NaOH	0.1 M HCl
0.2 M NaOH	0.2 M HCl
0.4 M NaOH	0.4 M HCl
0.8 M NaOH	0.8 M HCl

4. In this experiment, the indicator was present in each of the acidic samples. How would your procedure have changed if the indicator was present in just one of the acid samples?
5. If time permits, challenge a classmate to determine the identity of a set of unknowns that you provide him or her. Only one of the unknowns should have the indicator. Ask your teacher for the appropriate acid samples that do not contain any indicator to complete the set of eight unknowns.

WHERE HAVE ALL THE ELECTRONS GONE?

Introduction

In this experiment, two different types of oxidation-reduction reactions are investigated. In Part 1 of the experiment, an electrochemical cell is constructed using copper in a copper nitrate solution and zinc in a zinc nitrate solution. The effects of the oxidation-reduction reaction on the metal are used to determine how the electrons move in the cell. In Part 2 of the experiment, a steel paper clip is electroplated with copper.

Background Information

Chemical reactions that involve the transfer of electrons from one element to another are called oxidation-reduction reactions. The term oxidation-reduction refers to the coupled processes of oxidation-reduction. The process of oxidation-reduction occurs when a species gains an electron. A common example of an oxidation-reduction reaction is the rusting of iron. An example of an oxidation-reduction reaction that is more useful takes place inside a battery—or, more correctly, takes place when a battery is part of a completed circuit.

It may be difficult to visualize the transfer of electrons in an oxidation-reduction reaction. One method to measure the electron flow is to separate the oxidation half-reaction and the reduction half-reaction. These two half-reactions will not proceed unless the two are connected, as they are in an electrical circuit. Figure 1 shows the oxidation-reduction reaction between copper and iron separated into the individual half-reactions. The external wire connecting the iron and copper bars allows the electrons to flow from one half-reaction to the other. The salt bridge allows ions to move from one solution to the other to keep the charges balanced. In the reaction above, iron is oxidized to the iron (II) ion in solution. The electrons lost in this process travel through the wire to the copper, where copper (II) ions from the solution are reduced to copper metal. Since excess positive charge is building up in the iron solution (from the iron (II) ions) and positive charge is being removed from the copper solution, ions in the salt bridge move into the solutions to balance the charge. The iron electrode is called the anode; it is where oxidation-reduction takes place. The cathode is negatively charged.

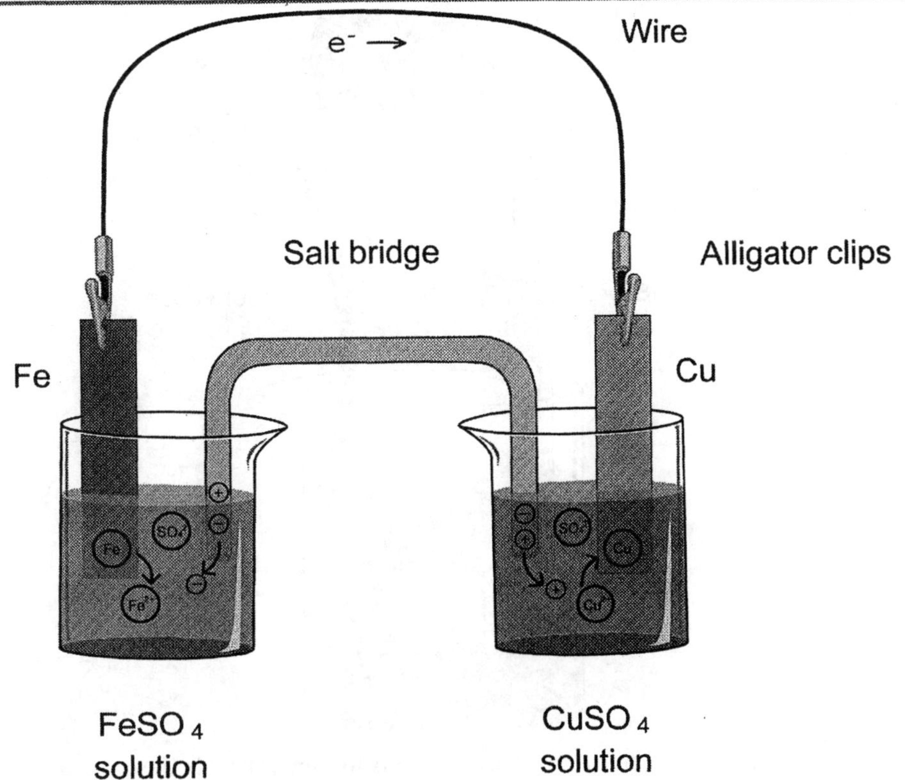

Figure 1. An electrochemical cell of the oxidation-reduction reaction between iron and copper.

The oxidation-reduction reaction between iron and copper in the example above occurs spontaneously; electrons naturally move from the iron electrode to the copper electrode. The industrial application of electroplating involves using electricity to force the electrons in the desired direction. Electroplating silver on top of copper wire is used to create high-conducting, long-lasting electrical connections. By connecting the negative end a of battery, or other electrical source, to the copper wiring, and connecting the positive end to a piece of silver, silver ions can be reduced to silver metal on top of the copper.

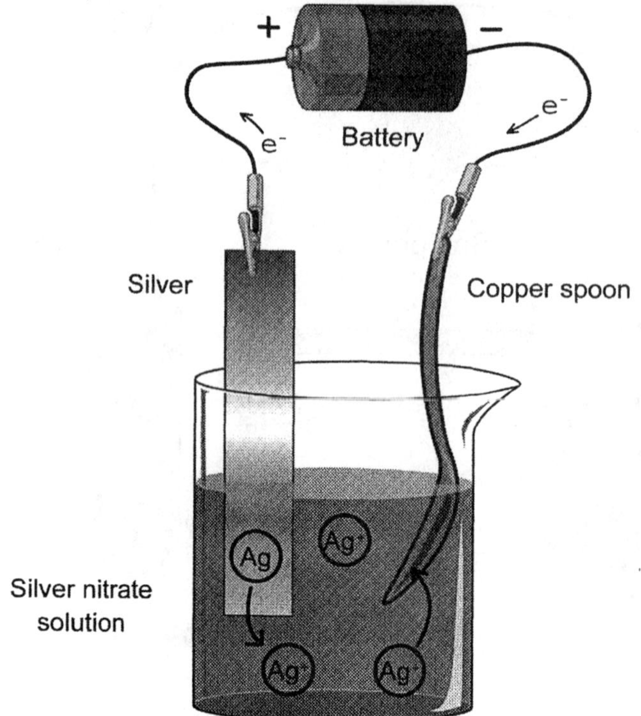

Figure 2. A battery forces electrons to flow towards the copper spoon in this simple electroplating setup.

In this scenario, the current provided by the battery must be large enough to overcome the natural flow of electrons from the copper to the silver. In order to determine the natural flow of electrons from one metal to another, use reference tables of standard reduction potentials.

The two experiments in this lab provide an opportunity to set up and explore oxidation-reduction reactions. The key to understanding the process is always knowing where the electrons will go.

Chemicals and Equipment

Part 1

- copper strip
- zinc strip
- two 250-mL beakers
- 0.1 M $CuNO_3$ solution
- 0.1 M $ZnNO_3$ solution
- saturated $NaNO_3$ (or KNO_3) solution
- large filter paper
- wire lead with alligator clips

Part 2

- 150-mL beaker
- 1 M $CuSO_4$ solution
- steel paper clip
- copper wire
- 6-volt battery
- two wire leads with alligator clips

Safety Concerns

Always wear safety goggles.

Procedure

Part 1

1. Prepare the zinc and copper strips by rubbing them with sandpaper to produce a shiny surface.
2. Weigh the zinc and copper strips. Make sure that the balance reads 0.000 g before placing the strips of metal on the balance. Record the mass to the nearest 0.001 g in the data table.
3. Add 150 mL of 0.1 M $CuNO_3$ solution to a 250-mL beaker. Place the copper strip in the solution.
4. Add 150 mL of 0.1 M $ZnNO_3$ solution to a second 250-mL beaker. Place the zinc strip in the solution.
5. From a piece of large filter paper, cut a 0.5″ wide strip across the middle of the filter paper. Soak the strip of filter paper in 0.1 M $NaNO_3$ solution. Once the paper is completely soaked, use tweezers to place the filter across the two beakers. Make sure each end of the filter is immersed in the solution.
6. Connect the two strips of metal using the wire lead with alligator clips.
7. After 20 minutes, carefully remove the metal strips from the solution and place them on a piece of paper towel. Gently dry the metal strips by pressing each side with paper towel.
8. You may want to let the strips dry while you complete the next part of the experiment. After they have thoroughly dried, weigh the zinc and copper strips. Make sure that the balance reads 0.000 g before placing the strips of metal on the balance. Record the mass to the nearest 0.001 g in the data table.

Part 2

9. Weigh a steel paper clip. Make sure that the balance reads 0.000 g before placing the paper clip on the balance. Record the mass to the nearest 0.001 g in the data table.
10. Fill a 150-mL beaker with 100 mL of $CuSO_4$ solution.
11. Attach the copper wire to the positive terminal of the battery using the wire lead with the alligator clips.
12. Attach the steel paper clip to the negative terminal of the battery using a second wire lead with alligator clips.

Important: Do not let the paper clips and the copper wire touch.

13. Carefully immerse the copper wire and paper clip in the $CuSO_4$ solution.
14. After 5 minutes, remove the paper clip from the solution and disconnect the alligator clips from the battery and the paper clip.
15. Gently dry the paper clip with paper towel. After the paper clip is dry, weigh it. Make sure that the balance reads 0.000 g before placing the paper clip on the balance. Record the mass to the nearest 0.001 g in the data table.
16. Using the same paper clip, reattach the wire leads and place the paper clip back in the solution.
17. After 5 minutes, remove the paper clip, gently dry it, and weigh it again. Make sure that the balance reads 0.000 g before placing the paper clip on the balance. Record the mass to the nearest 0.001 g in the data table.
18. Repeat steps 8 and 9.

Cleanup and Disposal

- Dispose of the copper nitrate and copper sulfate solutions in the appropriate waste container. Copper is toxic to plant and animal life and should not be disposed of in the sink.
- Dispose of the zinc nitrate solution in the appropriate waste container.
- The filter paper can be disposed of in the trash.
- The copper and zinc strips can be cleaned and reused.

Data and Observations

Part 1

Data table

Metal strip	copper	zinc
Initial mass of strip (g)		
Final mass of strip (g)		
Mass gained or lost (g)		

Part 2

Data table

Plating sessions	first	second	third
Final mass of paper clip (g)			
Initial mass of paper clip (g)			
Mass gained (g)			

Post Experiment Questions

1. Sketch a picture of the electrochemical cell in Part 1 of the experiment. Include the following labels: anode, cathode, and salt bridge. Use arrows to show the flow of electrons.
2. How much mass did the copper and zinc strips lose or gain in Part 1 of the experiment?
3. Using the amount of zinc lost in Part 1, how much copper should theoretically have been gained? Compare this amount to the actual amount of copper gained, as reported in the previous question.
4. What is the average weight of copper gained per minute in Part 2 of the experiment?
5. What is the percent by mass of copper on the paper clip at the end of the experiment? How many minutes would it take for the paper clip to be 50% copper?

WATER HARDNESS ANALYSIS

Introduction

The purpose of this experiment is to determine the hardness of a tap water sample by using a titration reaction. You will use ethylenediaminetetraacetic acid, EDTA, as the titrating agent. The indicator will be eriochrome black T, EBT, which is blue-colored when in a basic solution and pink-colored with in an acidic solution.

Background Information

Soap scum, bathtub rings, and clogged plumbing can often be blamed on water hardness. Water hardness is caused by the presence of significant concentrations of calcium ions, Ca^{2+}, and magnesium ions, Mg^{2+}. The presence of these ions will also prevent soaps from lathering. Ethylenediaminetetraacetic acid, EDTA, has the ability to bind to metal ions—like calcium and magnesium—and remain in solution.

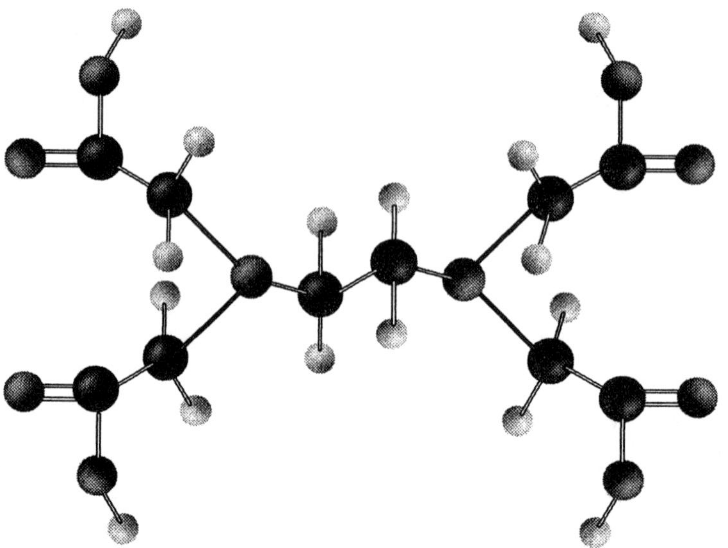

Figure 1. A ball-and-stick model of ethylenediaminetetraacetic acid.

EBT will be added to a water sample. This solution will be pink since the EBT forms complex ions with the dissolved Ca^{2+} or Mg^{2+} ions. As EDTA is added during the titration, the Ca^{2+} or Mg^{2+} ions will be released from the EBT and form a complex with EDTA according to the following equation:

$$Ca^{2+} + H_2(EDTA)^{2-} \rightarrow Ca(EDTA)^{2-} + 2H^+$$

EBT will have released the ions that are now complexed with the EDTA, and the solution will change color—from pink to blue—indicating the end point of the titration.

Chemicals and Equipment

- ~75 mL 0.0050 M EDTA solution
- ~15 drops EBT indicator solution
- ~30 mL pH=10 buffer solution of NH_3 and NH_4Cl
- deionized or distilled water
- 1 Erlenmeyer flask
- 10-mL graduated cylinder
- buret and buret clamp
- buret clamp
- ring stand
- 100-mL beaker
- dropper or beryl-style pipet
- 50-mL pipet
- pipet bulb or other delivery device

Safety Concerns

- **EDTA and the buffer solution are both strong bases. If by accident you get any on your skin, wash with copious amounts of water and notify your instructor immediately. If you accidentally spill any on your workspace, clean it up immediately. Vinegar, CH3COOH, should be used to neutralize any spilled base.**
- **The buffer solution contains ammonia, which produces a strong odor. Avoid breathing vapors.**
- **The EBT indicator is dissolved in alcohol and is therefore flammable. Avoid open flame.**
- **Goggles should be worn at all times.**

Procedure

1. Obtain a clean buret, buret clamp, and ring stand, and assemble these pieces of equipment. Place a clean Erlenmeyer flask under the buret.

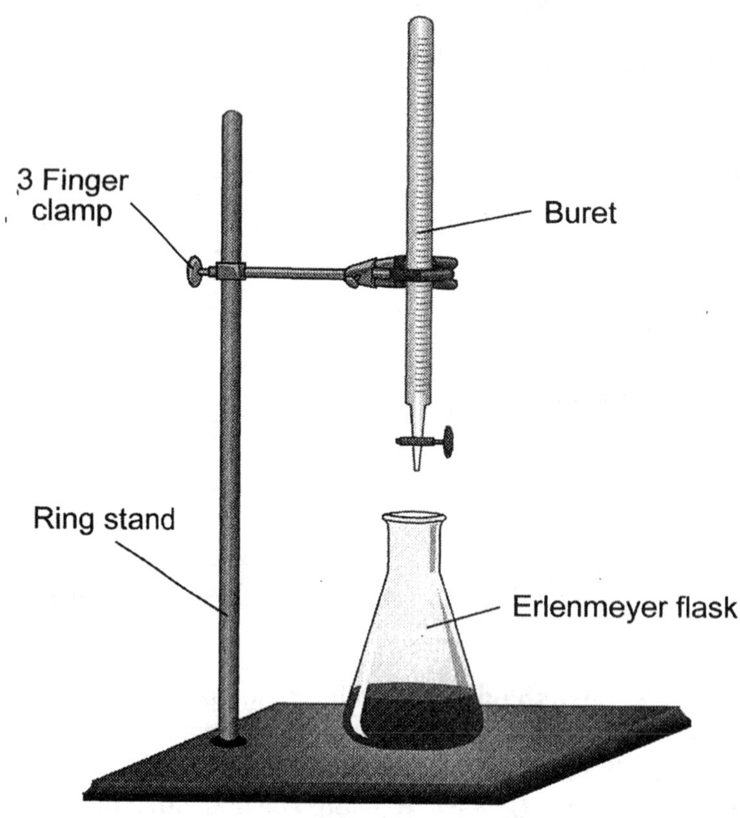

Figure 2. Titration assembly with buret.

2. Rinse the buret thoroughly with distilled or deionized water. Be sure to allow the distilled or deionized water to pass through the stopcock. Also be sure to rinse the tip of the stopcock. You do not want dissolved ions that are not in the water sample you are analyzing to interfere with the experiment.
3. Transfer approximately 75 mL of EDTA solution to a clean and dry 100-mL beaker.
4. Transfer 10 mL of the EDTA solution to the buret to use as a rinse. Drain the EDTA solution through the buret. Be sure to allow the EDTA solution to pass through the stopcock and be sure to rinse the tip of the stopcock.

5. Repeat the EDTA rinse.
6. Fill the buret with EDTA solution. Drain a small amount so that no air bubbles are present and so that the meniscus of the solution is on the graduated part of the buret. Record the starting position of the meniscus in your data table. Thoroughly clean the Erlenmeyer flask that you used to collect the EDTA. Be sure to rinse it with distilled or deionized water.

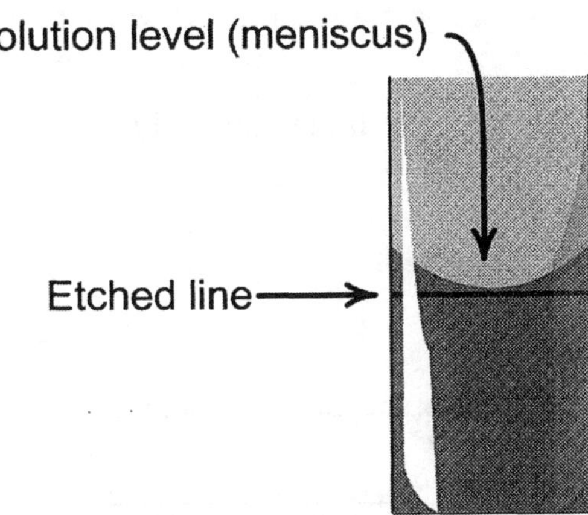

Figure 3. The bottom of the meniscus should be level with the etched line on the neck of the buret.

7. Using the Erlenmeyer flask that you just cleaned and rinsed with distilled or deionized water, pipet 50.0 mL of *tap* water into the flask.
8. Using a graduated cylinder, add 10 mL of pH=10 buffer solution to the water. Using a dropper, add 5 drops of EBT indicator solution.
9. Your water solution with buffer and EBT should be a pink color. If you have trouble seeing this color, place a piece of white paper beneath the Erlenmeyer flask as you begin the titration.
10. Place the Erlenmeyer flask under the buret and begin the titration. As you are adding EDTA from the buret to the Erlenmeyer flask, you will begin to see a purple color. When you begin to see this color change, you are nearing the end point of the titration. Continue to add EDTA, but do so drop by drop. When the solution changes to a sky-blue color, you have reached the end point and should stop adding EDTA. Record the final reading on the buret.
11. Wash the Erlenmeyer flask thoroughly and do a final rinse with distilled or deionized water. Repeat the titration two more times. Be sure to record your initial and final buret readings.

Cleanup and Disposal

- Be sure to wash thoroughly all of the equipment and supplies used during this lab.
- All solutions can be disposed of in the sink with generous amounts of water flushing down each sample before you add the next sample, unless otherwise directed by your instructor.
- Be sure to wash your hands thoroughly at the end of this lab.

Data and Observations

Data table

	Titration #1	Titration #2	Titration #3
Initial buret reading (mL)			
Finial buret reading (mL)			
EDTA solution used (mL)			

Averaged data (average the three titration values)
EDTA solution used: __________ mL

Post Experiment Questions

1. EDTA is a chelate. Define chelate. Show how EDTA works as a chelate, using its structure, which is found at the beginning of this lab.

The next few questions will help guide you in calculating the hardness of the water. Be sure to show all your calculations. Use the average values for these calculations.

2. Calculate the moles of EDTA required for the titration by using the molarity of the EDTA solution and the volume of EDTA necessary to reach the titration end point. Be sure to use the average volume of EDTA solution used.

$$\text{Molarity of EDTA solution} = \frac{\text{moles EDTA}}{\text{L EDTA solution}}$$

3. EDTA reacts in a 1:1 ratio with the ions that cause water hardness. This means that for every one mole of EDTA needed to reach the endpoint of the titration, one mole of ions was present. Calculate the molarity of the 50.0 mL water sample.

$$\text{Molarity of ions in solution} = \frac{\text{moles ions}}{\text{L water sample}}$$

Water hardness is typically reported in ppm Ca^{2+} (parts per million of Ca^{2+} ions). The next few questions will guide you through this calculation.

4. Convert the moles of ions to grams of ions by multiplying by the molecular weight of calcium. This will assume that all water hardness is caused by the presence of Ca^{2+} and not by other ions.

$$\text{grams } Ca^{2+} = (\text{moles } Ca^{2+}) \times (40.08 \text{ g/mole})$$

5. Calculate the ppm Ca^{2+} using the following formula. The grams of solution can be approximated as equal to the volume of the water sample since the density of water approximately 1.00 g/1.00 mL.

$$\text{ppm} \quad Ca^{2+} = \frac{\text{grams } Ca^{2+}}{\text{grams solution}} \times 1{,}000{,}000$$

6. Water that contains less than 120 ppm Ca^{2+} is considered to be soft water. Water that contains between 120 and 350 ppm Ca^{2+} is considered moderately hard and water that contains over 350 ppm Ca^{2+} is considered very hard. What would your water sample be considered?

Analysis of Bottled Water

Introduction

The objective in this experiment is to examine three aspects of bottled water. You will determine the hardness of the bottled water by following the same procedure that you followed determining the hardness of tap water. Review the discussion found in the water hardness lab prior to beginning this experiment. You will then determine the pH of the bottled water and, finally, you will determine the bicarbonate ion, HCO_3^-, concentration of the bottled water.

Background Information

An increasing number of Americans prefer bottled water to tap water. Since bottled water is approximately one thousand times more expensive than tap water, you may ask yourself why this phenomenon continues to grow. Many people report that bottled water is more pure and tastes better than tap water.

The presence of calcium ions, Ca^{2+}, and magnesium ions, Mg^{2+}, are the cause of water hardness, called **permanent hardness**. The presence of bicarbonate ions, HCO_3^-, is called **temporary hardness** and influences the pH of the water. The higher the bicarbonate ion concentration, the more alkaline the sample.

To determine the bicarbonate ion concentration, you will use HCl as a titrant and methyl orange as an indicator. As you add HCl to the bottled water, which contains bicarbonate ion, the HCl and bicarbonate ions react to produce water, carbon dioxide, and chloride ions according to the following equation:

$$HCO_3^- (aq) + HCl (aq) \rightarrow H_2O + CO_2 (g) + Cl^- (aq)$$

When all of the bicarbonate ions have reacted, adding additional HCl will only result in an excess of HCl, and the pH of the solution will drop. It is at this point that you will have reached the end point of the titration, and the methyl orange will change color. At the beginning of the titration the methyl orange will be yellow in color. When the pH drops below 4, the more acidic environment will cause the methyl orange to change from yellow to orange.

Some of your classmates will be analyzing a different brand of bottled water. Be sure to compare results before you leave.

Chemicals and Equipment

Bottled water hardness

- ~75 mL 0.0050 M EDTA solution
- ~15 drops EBT indicator solution
- ~30 mL pH=10 buffer solution of NH_3 and NH_4Cl
- deionized or distilled water
- Erlenmeyer flask
- 10-mL graduated cylinder
- buret
- buret clamp
- ring stand
- 100-mL beaker
- dropper or beryl-style pipet
- 50-mL pipet
- pipet bulb or other delivery device

pH measurement

- pH paper
- dropper or beryl-style pipet

Bicarbonate ion measurement

- 24 well reaction plate
- two 50-mL beakers
- two beryl-style pipets
- small stirring device (to be used with the reaction plate)
- ~20-mL bottled water

- ~0.020 M HCl *[This will be standardized by the lab instructor. Be sure to record its exact concentration in your laboratory manual.]*
- ~5 drops methyl orange indicator

Safety Concerns

- **EDTA and the buffer solution are both strong bases. If by accident you get any on your skin, wash with copious amounts of water and notify your instructor immediately. If you accidentally spill any on your workspace, clean it up immediately. Vinegar, CH_3COOH, should be used to neutralize any spilled base.**
- **The buffer solution contains ammonia, which produces a strong odor. Avoid breathing vapors.**
- **The EBT indicator is dissolved in alcohol and is therefore flammable. Avoid open flame.**
- **Although the HCl solution has a low concentration, hydrochloric acid is nonetheless a strong acid. You should follow careful safety procedures when working with it. If you accidentally spill any on yourself or your workspace, baking soda may be used to neutralize the spill. Wash with copious amounts of water and notify your instructor immediately.**
- **Goggles should be worn at all times.**

Procedure

Bottled-water hardness

1. Obtain a clean buret, buret clamp, and ring stand, and assemble these pieces of equipment. Place a clean Erlenmeyer flask under the buret.
2. Rinse the buret thoroughly with distilled or deionized water. Be sure to allow the distilled or deionized water to pass through the stopcock. Also be sure to rinse the tip of the stopcock. You do not want dissolved ions that are not in the water sample you are analyzing to interfere with the experiment.
3. Transfer approximately 75 mL of EDTA solution to a clean and dry 100-mL beaker.
4. Transfer 10 mL of the EDTA solution to the buret to use as a rinse. Drain the EDTA solution through the buret. Be sure to allow the EDTA solution to pass through the stopcock and be sure to rinse the tip of the stopcock.
5. Repeat the EDTA rinse.

6. Fill the buret with EDTA solution. Drain a small amount so that no air bubbles are present and so that the meniscus of the solution is on the graduated part of the buret. Record the starting position of the meniscus in your data table. Thoroughly clean the Erlenmeyer flask that you used to collect the EDTA. Be sure to rinse it with distilled or deionized water.
7. Using the Erlenmeyer flask that you just cleaned and rinsed with distilled or deionized water, pipet 50.0 mL of *bottled* water into the flask.
8. Using a graduated cylinder, add 10 mL of pH=10 buffer solution to the water. Using a dropper, add 5 drops of EBT indicator solution.
9. Your water solution with buffer and EBT should be a pink color. If you have trouble seeing this color, place a piece of white paper beneath the Erlenmeyer flask as you begin the titration.
10. Place the Erlenmeyer flask under the buret and begin the titration. As you are adding EDTA from the buret to the Erlenmeyer flask you will begin to see a purple color. When you begin to see this color change, you are nearing the end point of the titration. Continue to add EDTA, but do so drop by drop. When the solution changes to a sky-blue color, you have reached the end point and should stop adding EDTA. Record the final reading on the buret.
11. Thoroughly wash out the Erlenmeyer flask and do a final rinse with distilled or deionized water. Repeat the titration two more times. Be sure to record your initial and final buret readings.

pH measurement

12. Using a dropper or beryl-style pipet, place a drop of bottled water on a piece of pH paper. Compare the color of the pH paper to the standard colors. Record the pH value in your data table.

Bicarbonate ion measurement

You have already performed a more standard titration as part of this laboratory experiment. To determine the bicarbonate ion concentration, you will use a microscale titration. Before you begin this part of the experiment, review the chemistry behind it, which is briefly discussed at the beginning.

13. Obtain a clean, dry 24-well reaction plate. If your laboratory bench surface is dark, place a piece of plain white paper under the reaction plate. This will make it easier for you to see the methyl orange indicator change from yellow to orange as the reaction approaches the end point and the pH drops below 4.

14. Transfer ~10 mL of HCl solution to a 50-mL beaker. The HCl solution has been standardized by your instructor. This is necessary because it is essential to know the exact concentration of the HCl solution. Standardization is the method by which the exact concentration is determined. Your instructor will have reported the exact concentration. Label your beaker with this exact concentration and record this value in the data table.

15. Transfer ~20 mL bottled water to a 50-mL beaker. Label your beaker with the brand of water and record this information in the data table.

16. You will be adding HCl and water to the well plate using a beryl-style pipet. It is crucial that you do not confuse the identity of the two pipets; therefore, label them. Label one as "HCl" and the other as "H_2O". A permanent marker should work, as should a wax pencil.

Yellow to orange is a difficult color change to observe. Therefore, the next part of this titration is designed to help you with your observations.

17. Add 10 drops of bottled water to one well in your reaction plate.

18. Add 1 drop of methyl orange indicator. The solution should now turn yellow.

19. Carefully and slowly begin adding drops of HCl to this well. Using a small stirring device, stir between each drop. Count the number of HCl drops you need to add until an orange color persists. Record the number of drops in the data table. This number will not be used in a calculation, but will serve as a target number for subsequent titrations.

20. To go back to the original yellow color, add enough drops of bottled water for this color change to occur.

21. Thoroughly wash and dry the stirring device.

Qualitative titrations

22. To a new well on your reaction plate, add 10 drops of bottled water and 1 drop of methyl orange. The resulting solution should be yellow.

23. Carefully and slowly begin adding drops of HCl. Be sure to stir between drops. Count the number of drops of HCl that you need to add until an orange color persists. Record this number in the data table.

24. Thoroughly wash and dry the stirring device.

25. Repeat this procedure two more times.

Cleanup and Disposal

- Be sure to wash thoroughly all of the equipment and supplies used during this lab.
- All solutions can be disposed of in the sink with generous amounts of water flushing down each sample before you add the next sample, unless otherwise directed by your instructor.
- Be sure to wash your hands thoroughly at the end of this lab.

Data and Observations

Brand of bottled water: ______________________________

Bottled water hardness

	Titration #1	Titration #2	Titration #3
Initial buret reading (mL)			
Final buret reading (mL)			
EDTA solution used (mL)			

Averaged data (average the three titration values)
EDTA solution used __________ mL

pH measurement

pH of bottled water: __________

Bicarbonate ion measurement

__________ M HCl exact concentration

Reference titration

__________drops of HCl added

Qualitative titrations

	Titration #1	Titration #2	Titration #3
Drops of HCl added			

Averaged data (average the three titration values)

__________ drops of HCl added

Post Experiment Questions

The next few questions will help guide you in calculating the hardness of the water. Be sure to show all your calculations. Use the average values for these calculations.

1. Calculate the moles of EDTA required for the titration by using the molarity of the EDTA solution and the volume of EDTA necessary to reach the titration end point. Be sure to use the average volume of EDTA solution used.

$$\text{Molarity of EDTA solution} = \frac{\text{moles EDTA}}{\text{L EDTA solution}}$$

2. EDTA reacts in a 1:1 ratio with the ions that cause water hardness. This means that for every one mole of EDTA needed to reach the endpoint of the titration, one mole of ions was present. Calculate the molarity of the 50.0 mL water sample.

$$\text{Molarity of ions in solution} = \frac{\text{moles ions}}{\text{L water sample}}$$

Water hardness is typically reported in ppm Ca^{2+} (parts per million of Ca^{2+} ions). The next few questions will guide you through this calculation.

3. Convert the moles of ions to grams of ions by multiplying by the molecular weight of calcium. This will assume that all water hardness is caused by the presence of Ca^{2+} and not by other ions.

$$\text{grams } Ca^{2+} = (\text{moles } Ca^{2+}) \times (40.08 \text{ g/mole})$$

4. Calculate the ppm Ca^{2+} using the following formula. The grams of solution can be approximated as equal to the volume of the water sample since the density of water approximately 1.00 g/1.00 mL.

$$\text{ppm } Ca^{2+} = \frac{\text{grams } Ca^{2+}}{\text{grams solution}} \times 1{,}000{,}000$$

5. Water that contains less than 120 ppm Ca^{2+} is considered to be soft water. Water that contains between 120 and 350 ppm Ca^{2+} is considered moderately hard and water that contains over 350 ppm Ca^{2+} is considered very hard. What would your water sample be considered?

The next few questions will help guide you in calculating the concentration of bicarbonate in your bottled water sample. Be sure to show all your calculations. Use the average values for these calculations.

6. Calculate the molarity of bicarbonate ion, HCO_3^-, using the following formula:

$$HCO_3^- \text{ molarity} = \text{HCl molarity} \times \frac{\text{drops of HCl}}{\text{drops of bottled water}}$$

7. As stated before, ion concentration is typically reported as parts per million. To calculate the ppm HCO_3^-, use the following equation:

$$\text{ppm } HCO_3^- = (HCO_3^- \text{ molarity}) \times (0.061) \times (1{,}000{,}000)$$

8. Compare your pH value and ppm HCO_3^- with those of your classmates who used a different brand of bottled water. Are there differences? What type of relationship exists between the pH and the bicarbonate ion concentration?

Measurements of Particulates in Smoke

Introduction

The goal of this experiment is to capture and measure the particulates contained in smoke. A cigarette and a mini-cigar will provide a source of smoke in this experiment.

Background Information

Smoke consists of **particulate matter**, or **particulates** for short. Particulates are small solid or liquid particles suspended in air. These small particles would be invisible to the eye if viewed one by one; however, when collected as a group, particulates appear as a cloud or haze. The amount of particulates in the air is one of the qualities used to measure air quality. High concentrations of particulate matter are known to cause, or increase the severity of, many breathing related conditions. Particulates are classified as either coarse (larger than 2.5 μm) or fine (smaller than 2.5 μm). Both coarse and fine particulates pose health risks, though the esophagus and hairs in the nose filter some coarse particulates from reaching the lungs.

The smoke produced when tobacco leaves are burned contains thousands of components. The gaseous components include carbon monoxide, nitrogen dioxide, cadmium, polycyclic aromatic hydrocarbons, formaldehyde, and other volatile organic carbon (VOC) compounds. The smoke also contains small ash particles (called flyash), tar particles, and other pieces of dust. When these small particles are inhaled, they enter the lungs. Once inside the lungs, the small particles can irritate and destroy the lung cells.

Research has shown that a smoker is ten times more likely to develop lung cancer than a nonsmoker. Workers in occupations where smoke or dust inhalation is common also are more likely to develop emphysema and lung cancer than other workers. Secondhand smoke, or smoke that is exhaled by smokers or that emanates from a lit cigarette or cigar, has been listed in the same category as other airborne carcinogens, such as asbestos.

In this experiment, you will measure the particulate matter in the smoke from a cigarette and from a mini-cigar or clove cigarette. The smoke inhaled by the smoker will be analyzed, as will the smoke that escapes into the environment. A vacuum will draw the smoke through a piece of filter paper where the particulates will be trapped. Capturing for analysis the smoke released into the environment will prevent you from inhaling the smoke and pose a minimal risk to you.

Chemicals and Equipment

- two 250-mL vacuum flasks
- two rubber stoppers with a glass tube
- rubber tubing
- two glass funnels
- ring stand with 2 small three-finger clamps
- filter paper (capable of filtering 2.5 μm particles)
- cigarettes (filtered)
- mini-cigars or cloves

Safety Concerns

Always wear safety goggles.

Procedure

1. The assembly has two components, a smoker flask and an environment flask.

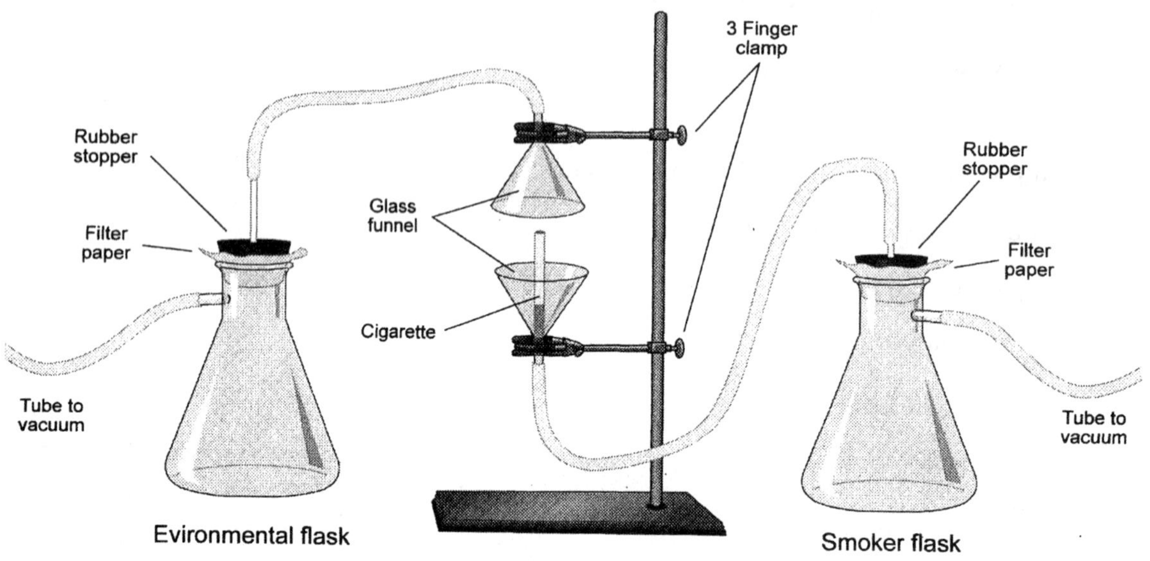

Figure 1. Smoke collection assembly

2. The smoker flask is a 250-mL vacuum flask connected to a glass funnel with rubber hose. The cigarette is placed in the glass funnel. When the cigarette is lit and the vacuum turned on, smoke is drawn through the cigarette and into the smoker flask, where the particles are trapped on the filter paper.
3. The environment flask is a 250-mL vacuum flask connected to a glass funnel with rubber hose. The funnel is placed about 1 cm above the cigarette. When the vacuum is turned on, smoke from the cigarette is drawn into the environment flask and trapped on the filter paper.
4. To reduce smoke escaping into the atmosphere, make sure that the equipment is assembled and functions properly before you light the cigarette.
5. Remove both pieces of filter paper from the assembly and weigh them to the nearest 0.001 g. Make sure the balance reads 0.000 g before placing the filter paper on the balance. Record the masses in the data table.
6. Place the filter paper back in the assembly.
7. Prepare to light the cigarette by turning on the smoker-flask vacuum.
8. Light the cigarette and immediately turn on the environment-flask vacuum.
9. Adjust the vacuum on the smoker flask so that the cigarette has burned down to the last centimeter in about 4 minutes.

10. Put out the cigarette with a drop of water and turn off the vacuums to both flasks.
11. Remove the filter paper from both flasks and weigh them to the nearest 0.001 g. Make sure the balance reads 0.000 g before placing the filter paper on the balance. Record the masses in the data table.
12. Remove the cigarette and accumulated ash from the funnel of the smoker flask. Dispose of the waste as directed by your instructor.
13. Repeat steps 2 – 9 with a new set of filter paper and cigarette. To mimic actual smoking, increase the vacuum to the smoker flask and pinch the rubber tubing. Set a regimen of inhalation time (unpinched tubing) to rest time (pinched tubing). Record your regimen in the data table.
14. Repeat steps 2 – 10 using a mini-cigar or clove cigarette.

Cleanup and Disposal

- All equipment should be washed thoroughly with warm soapy water.

Data and Observations

Data table

	Cigarette		Mini-cigar or clove	
Trial	Constant inhalation	Realistic inhalation	Constant inhalation	Realistic inhalation
Smoking regimen	N/A		N/A	
Mass of dirty smoker filter paper (g)				
Mass of clean smoker filter paper (g)				
Mass of particulate matter inhaled by smoker (g)				
Mass of dirty environment filter paper (g)				
Mass of clean environment filter paper (g)				
Mass of particulate matter released to the environment (g)				

Post Experiment Questions

1. If a smoker smokes a pack of cigarettes a day, how much particulate matter will be inhaled by the smoker and how much will be released to the environment? (There are 20 cigarettes in a pack of cigarettes.)
2. What affect does mimicking actual smoking have on the results?
3. Which product—mini-cigars or cigarettes—poses a more significant danger to the smoker and to the surrounding environment?
4. What prevents us from using the data from this experiment to determine the amount of particulate matter inhaled through secondhand smoke?
5. How effective are filtered cigarettes in removing particulate matter from the smoke inhaled by the smoker? Design an experiment to determine the answer and, if time permits, carry out the experiment.

Ultraviolet Light Detection

Introduction

The purpose of this experiment is to analyze the protective qualities provided by different substances from ultraviolet light. You will use beads sensitive to ultraviolet light as an indicator to measure the amount of ultraviolet light protection.

Background Information

Compared to visible light, ultraviolet light is at a higher energy level and, in addition to many other adverse effects, has been shown to cause skin cancer by causing changes in DNA. Earth has a protective layer of ozone, O_3, which helps to filter out ultraviolet light. However, due to the chemicals we are using and producing on Earth, we are causing a depletion of the ozone layer. As the ozone layer is being depleted, more and more ultraviolet light is reaching Earth's surface, and we need to be increasingly aware of our exposure.

People use all sorts of devices and chemicals to protect themselves from UV-light exposure. Some sunglasses have lenses that absorb either UV-A, which has a wavelength of between 320 nm and 400 nm, or UV-B, which has a wavelength of between 280 nm and 320 nm. People who are exposed to UV light outdoors may wear long-sleeved clothes for protection or apply some sort of lotion. Sunbathers often apply baby oil or a lotion with a sun protection factor (SPF).

SPF lotions work by either absorbing the UV light or blocking and scattering the UV light. SPF lotions only protect us from UV-B exposure. No lotion has been shown to significantly protect us from UV-A exposure. UV-A light exposure is a cause of skin aging and may increase the risk of malignant melanoma, which is the most dangerous form of skin cancer. UV-B light exposure increases the risk of basal cell carcinoma and squamous-cell carcinoma, which are two forms of non-melanoma skin cancer. UV-B is also responsible for sunburn.

The SPF values reported on sunscreen lotions are measured by timing how long skin samples covered with a sunscreen lotion take to burn compared to unprotected skin samples. For example, a sunscreen lotion that reports it has an SPF of 15 means that skin covered with the lotion takes 15 times longer to burn than if it was unprotected. The actual minutes depend on your skin color and tone.

The point of this experiment is for you to explore various materials and lotions and compare how well they protect things from UV light. You will be given plastic beads that have been impregnated with photosensitive chemicals. When exposed to an ultraviolet light source, they will change from white to a variety of colors. The more UV light that the beads are exposed to, the faster they change color. The beads will return to

white when placed in the dark. You will test the protective ability of plastic wrap, wax paper, plain white paper, baby oil, and a thin white or unbleached fabric, as well as two sunscreens, one with an SPF of 15 and another with an SPF of 45. By measuring the time it takes for the beads to change color, you will be measuring the protective value of the various materials and chemicals.

Chemicals and Equipment

- 24 UV-sensitive beads
- sunscreen with SPF 15
- sunscreen with SPF 45
- baby oil
- wax paper
- plastic wrap
- plain white paper
- thin white or unbleached cloth
- 8 plastic or paper weighing boats
- timing device with a second hand
- sunlight or an artificial source of ultraviolet light

Safety Concerns

- **If you are using an artificial ultraviolet light as your UV source, do not directly expose your eyes and do protect your skin.**
- **Although the sunscreens are considered safe, they can be greasy, therefore be careful when applying them to the beads.**

Procedure

Please note: This experiment does not need to be conducted inside a laboratory room using an artificial source of UV light. It can be done effectively outside, using the sun as the UV light source.

1. You will need to obtain three beads for each protective material or chemical. You also need an additional three beads that will serve as your experimental control.
2. Once you have gathered the 24 beads and the seven protective materials or substances, your task is to prepare an experimental surface. Obtain eight plastic or paper weighing boats. Each boat should be labeled with the protective material or substance to be used. One will be labeled as the experimental control.
3. Prepare each set of three beads for testing before exposing any of the beads to UV light.
4. Thoroughly coat a set of three beads, inside and outside, with baby oil. You do not want the beads to be dripping with baby oil, only thoroughly coated. Place them in their labeled weighing boat.
5. Thoroughly coat a set of three beads, inside and outside, with an SPF 15 sunscreen lotion. Again you do not want the beads to have an overly thick coating of sunscreen lotion, but they do need to be thoroughly coated. Place them in their labeled weighing boat.
6. Thoroughly coat a set of three beads, inside and outside, with an SPF 45 sunscreen lotion. Again you do not want the beads to have an overly thick coating of sunscreen lotion, but they do need to be thoroughly coated. Place them in their labeled weighing boat.
7. Wrap three beads in plastic wrap and place them in their labeled weighing boat.
8. Wrap three beads in wax paper and place them in their labeled weighing boat.
9. Wrap three beads in plain white paper and place them in their labeled weighing boat.
10. Wrap three beads in thin white or unbleached fabric and place them in their weighing boat.
11. Place three unprotected beads in their labeled weighing boat. These will serve as your experimental control.
12. Take your weighing boats and your timing device to the area where they will be exposed to UV light. You can either expose all of the samples at the same time or expose them one at a time.
13. Begin timing as soon as the exposure begins and stop timing as soon as you definitely notice a change in color. The colors are not dramatic, but rather a milky color.

14. Record the times and any observations on the data chart.

Cleanup and Disposal

- ☞ Wash your beads thoroughly, being careful not to lose any down the drain. Dry them off and return them to your instructor.
- ☞ Return all materials to your instructor.
- ☞ Be sure to wash your hands thoroughly and clean up your laboratory space.

Data and Observations

General Observations:

Data table

Test	Time (seconds)
Experimental control	
Plastic wrap	
Wax paper	
Plain white paper	
Fabric	
Baby oil	
SPF 15 sunscreen	
SPF 45 sunscreen	

Post Experiment Questions

1. What conclusion(s) have you reached?
2. In addition to UV-A and UV-B, there is also UV-C. UV-C has a wavelength range of between 200 nm and 280 nm. What is a practical use of UV-C?
3. Many flowers display a spectral signature that helps bees and insects pollinate the flower. This spectral signature is visible using a UV light source. What are some of the flowers that display this type of spectral signature, and what are some of the common patterns that are seen on the flowers?
4. Although exposure to UV light is considered harmful, what are two good uses for UV light? What types of UV light are used for these two good uses?
5. Why were you required to include an experimental control? What is the purpose of including an experimental control?
6. During what time of the day would the UV-sensitive beads change their color the fastest? Why?
7. It was suggested that your experimental surface be a plastic or paper weighing boat. These are usually white in color. Would your results have changed if you had used something that was colored or black? Why or why not?

ANALYSIS OF GAS SAMPLES

Introduction

The goal of this experiment is to analyze four different gas samples for their acidity and reactivity. First you will collect four different gas samples. You will generate the first gas sample, carbon dioxide, CO_2, by reacting sodium bicarbonate, $NaHCO_3$, with hydrochloric acid, HCl. You will generate the second gas sample, oxygen, O_2, by breaking down hydrogen peroxide, H_2O_2. The third gas sample will be exhaled breath, which you will capture in a zippered bag. The fourth gas sample will be ordinary air. by reacting each with bromthymol blue indicator solution to test the acidic or basic nature of the gas sample, and to test the reactivity of each gas sample with lime water, which is a saturated aqueous calcium hydroxide solution.

Background Information

Bromthymol blue indicator solution will be used to study the acidic or basic nature of the four gas samples. In an acidic environment, bromthymol blue is a yellow color. In a basic environment, it is a blue color. You are not determining a quantitative pH, but rather you are making comparative observations and developing qualitative data.

You will also test each of the four gas samples by reacting each with a saturated aqueous solution of calcium hydroxide, which is commonly called limewater. When $Ca(OH)_2$ (aq) is in the presence of CO_2 (g), the solution will appear milky due to the precipitation of calcium carbonate, $CaCO_3$, according to the following reaction:

$$Ca(OH)_2\ (aq) + CO_2\ (g) \rightarrow CaCO_3\ (s) + H_2O$$

Once again, you are not making quantitative measurements. You are making qualitative observations.

Chemicals and Equipment

- 1 teaspoon or ~2.0 grams of sodium bicarbonate, $NaHCO_3$
- a pinch or 0.5 gram of potassium iodide, KI
- ~10 ml 10% hydrogen peroxide, H_2O_2
- ~10 mL 20% aqueous hydrochloric acid, HCl
- ~ 2mL bromthymol blue indicator solution
- ~2 mL standard limewater solution, saturated aqueous $Ca(OH)_2$ solution
- plain piece of white paper

- piece of black paper (not necessary if the lab bench is black)
- 3 heavy-duty or freezer pint-sized zipper storage bags
- 10 plastic beryl-style pipets
- 24-well reaction plate
- 2 plastic teaspoons
- 1 plastic straw

Safety Concerns

- **The 10% hydrogen peroxide, H_2O_2, and the 20% hydrochloric acid, HCl, are both corrosive solutions. You should handle both with prudence. If by accident you get any on your skin, wash with copious amounts of water and notify your instructor immediately. If you accidentally spill any on your workspace, clean it up immediately.**
- **Goggles should be worn at all times.**
- **Safety gloves should be worn when working with H_2O_2 and HCl.**

Procedure

Preparation of a CO_2 sample

1. Place ~2 grams (a teaspoon) of sodium bicarbonate, $NaHCO_3$, in the bottom corner of a heavy-duty or freezer zipper storage bag that has been labeled "CO_2."
2. Completely fill a beryl-style pipet with 20% hydrochloric acid, HCl, solution. To help ensure that you completely fill the pipet, begin by squeezing all of the air out and then fill the bulb with the acid.
3. Place the acid-filled pipet in the zipper storage bag just above the $NaHCO_3$ sample. Carefully smooth the bag out, without releasing the acid, so that nearly all of the air is removed. Carefully and quickly zip the storage bag entirely closed.
4. Hold the sealed zipper storage bag by the corner opposite the $NaHCO_3$ sample. Slowly squeeze the pipet so as to release the acid onto the $NaHCO_3$ sample.

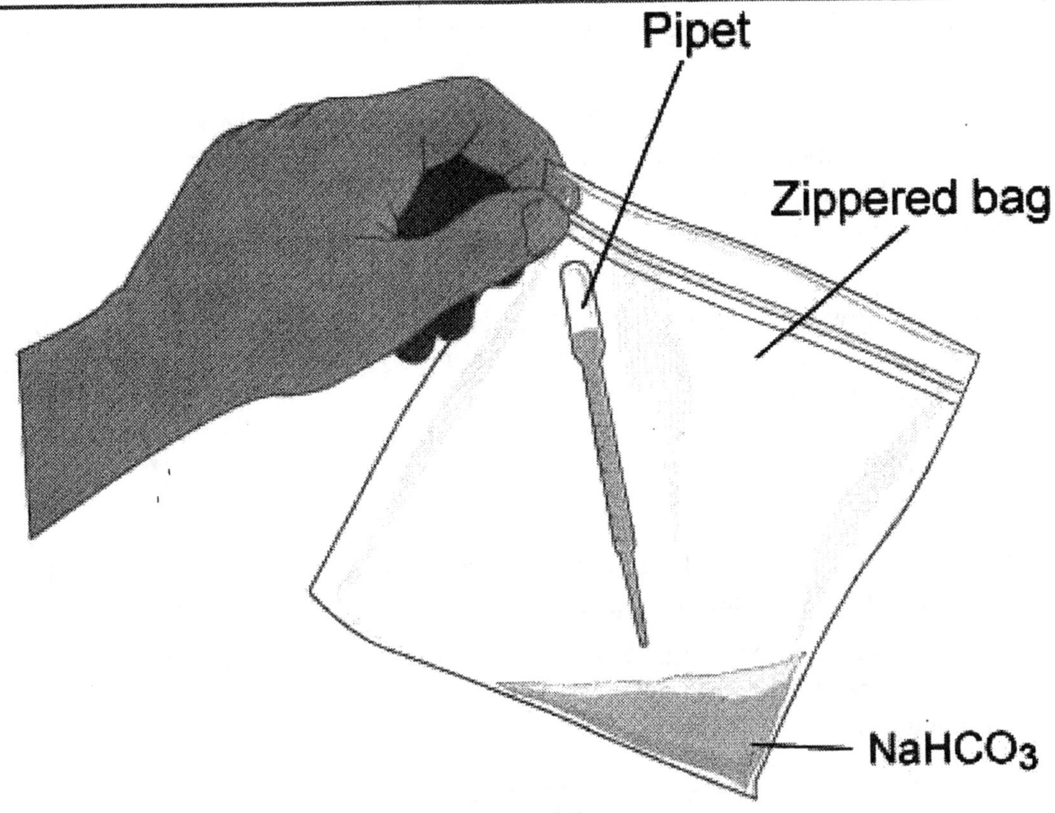

Figure 1. A pipet filled with 20% HCl is sealed in a zippered bag with $NaHCO_3$.

5. A reaction will begin. The bag should remain sealed throughout this process. Do not open it until you are directed to do so. Observe what happens throughout the duration of the reaction. Record these observations in the data table. Do your observations correspond to the reaction as described in the introduction for this laboratory experiment?

6. Your bag now contains a sample of carbon dioxide gas, CO_2. You will use this sample in the three tests that follow. Do not open the bag until you are directed to do so. Prop it up upright against the laboratory bench.

Preparation of an O_2 sample

7. Place a small pinch (less than 0.5 grams) of potassium iodide, KI, in the bottom corner of a heavy duty or freezer zipper storage bag that has been labeled "O_2."

8. Completely fill a beryl-style pipet with 10% hydrogen peroxide, H_2O_2, solution. To help ensure that you completely fill the pipet, begin by squeezing all of the air out and then fill the bulb with the hydrogen peroxide.

9. Place the peroxide-filled pipet in the zipper storage bag just above the KI sample. Carefully smooth the bag out, without releasing the hydrogen peroxide, so that

nearly all of the air is removed. Carefully and quickly zip the storage bag entirely closed.

10. Hold the sealed zipper storage bag by the corner opposite the KI sample. Slowly squeeze the pipet so as to release the H_2O_2 onto the KI sample. A reaction will begin. The bag should remain sealed throughout this process. Do not open it until you are directed to do so. Observe what happens throughout the duration of the reaction. Record these observations in the data table.
11. Your bag now contains a sample of oxygen gas, O_2. You will use this sample in the tests that follow. Do not open the bag until you are directed to do so. Prop it up upright against the laboratory bench.

Capture of breath

12. Insert a plastic straw into a third zippered storage bag that you have labeled "breath." Partially seal the bag shut so that only the part with the straw is open.
13. Deeply inhale and exhale into the straw to fill the storage bag with your breath. You may need to repeat this process several times to ensure that the bag is completely filled.

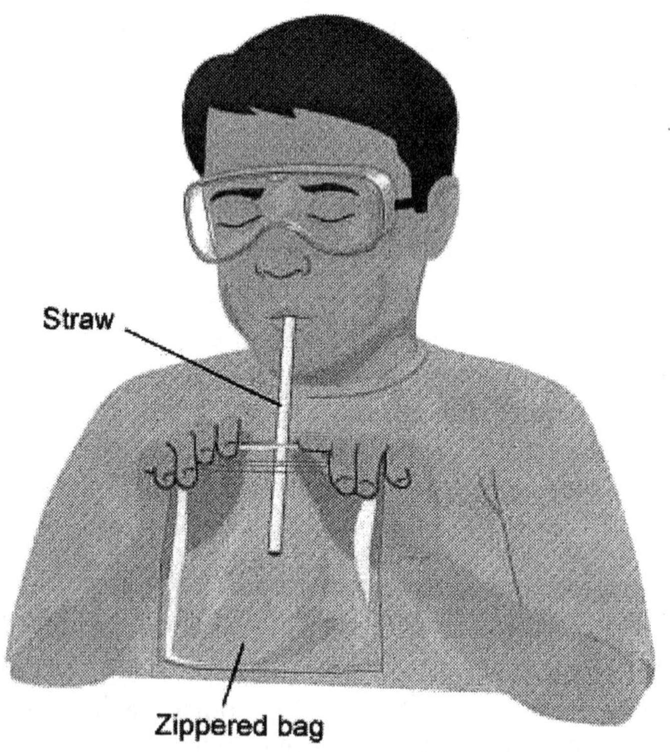

Figure 2. Seal the bag around the straw while capturing your breath.

14. Quickly remove the straw and seal the storage bag completely.
15. Your bag now contains a sample of your breath, which should contain CO_2. You will use this sample in the tests that follow. Do not open the bag until you are directed to do so. Prop it up upright against the laboratory bench.

How to obtain a gas sample for testing

For each of the following tests, you will need to use a full pipet of gas. For each test or gas source you will need to use a clean, fresh pipet.

16. To help ensure that you completely fill the pipet with a gas sample, begin by squeezing all of the air out of it.
17. Quickly and carefully insert the pipet into the zipper storage bag at a spot where you have just barely opened the seal. Be careful to not touch the tip of the pipet to any of the materials used to generate the gas sample.
18. Release the bulb of the pipet so that it fills with the gas sample.
19. Quickly and carefully remove the pipet and immediately reseal the zipper storage bag.
20. To obtain an ordinary air sample, simply hold a pipet up above your head, squeeze the tip, release it, and the pipet will fill with ordinary air.

Acidity test

You will test CO_2, O_2, exhaled breath, and a sample of ordinary air for their reactions with bromthymol blue indicator, which is an acid/base indicator. In a basic environment, bromthymol blue is blue in color and in an acidic environment is it yellow in color. Although you will not determine exact pH values, you will compare the results of the four gas samples tested.

21. Obtain a clean 24-well reaction plate. If your laboratory bench is dark in color, you should place a piece of plain white paper beneath it so that you can detect color changes more easily.
22. Add 10 drops of bromthymol blue indicator solution to the first well.
23. Obtain a sample of O_2 gas in a clean pipet, as outlined in the directions above.
24. Carefully and slowly bubble the gas sample through the bromthymol blue indicator solution by placing the tip into the well and gently squeezing the tip of the pipet. Record your observations in the data table.
25. Repeat the previous three steps for a CO_2 gas sample, an exhaled breath sample, and an ordinary air sample. Record all your observations in the data table.

Reaction with calcium hydroxide solution, $Ca(OH)_2$

You will now test each of the four gases by reacting them with an aqueous solution of calcium hydroxide, which is commonly called limewater. When $Ca(OH)_2$ (aq) is in the presence of CO_2 (g), the solution will appear milky due to the precipitation of calcium carbonate, $CaCO_3$ according to the following reaction:

$$Ca(OH)_2\ (aq) + CO_2\ (g) \rightarrow CaCO_3\ (s) + H_2O$$

Once again, you are not making quantitative measurements. You are making qualitative or comparative observations. You will use the sample reaction plate as before; simply move down to another line.

26. Place the well plate on a dark surface so that you can easily see the formation of the milky precipitate.
27. Add 10 drops of limewater solution to the first well.
28. Obtain a sample of O_2 gas in a clean pipet, as outlined in the directions above.
29. Carefully and slowly bubble the gas sample through the bromthymol blue indicator solution by placing the tip into the well and gently squeezing the tip of the pipet. Record your observations in the data table.
30. Repeat the previous three steps for a CO_2 gas sample, an exhaled breath sample, and an ordinary air sample. Record all of your observations in the data table.

Cleanup and Disposal

- Be sure to wash thoroughly all of the equipment and supplies used during this lab. Pipets and the reaction plate can be reused if thoroughly washed and rinsed with distilled or deionized water.
- All solutions can be disposed of in the sink with generous amounts of water flushing down each sample before you add the next sample, unless otherwise directed by your instructor.
- Zipper bags can be disposed of in the trash can once they have been thoroughly rinsed out.
- Be sure to wash your hands thoroughly at the end of this lab.

Data and Observations

General observation for the generation of CO_2

General observations for the generation of O_2

Bromthymol blue tests

General observations and conclusions:

Data table

Gas sample	Observation
CO_2	
O_2	
exhaled breath	
ordinary air	

Limewater tests

General observations and conclusions:

Data Table

Gas sample	Observation
CO_2	
O_2	
exhaled breath	
ordinary air	

Post Experiment Questions

1. What conclusions have you reached concerning the acidic or basic nature of the reactions that you observed?
2. What conclusion have you reached concerning the reaction of limewater with the four gas samples?
3. Which contains more CO_2, exhaled breath or ordinary air? Why?
4. Potassium iodide, KI, serves as a catalyst for the generation of O_2 from H_2O_2. What is a catalyst?
5. Write a balanced equation for the reaction of sodium bicarbonate, $NaHCO_3$, and hydrochloric acid, HCl.

Solar-Assisted Environmental Remediation

Introduction

The objective in this experiment is to determine the roles sunlight and titanium dioxide[1] play in the remediation of water contaminated with methylene blue. The breakdown of contaminants in polluted waters is a complicated process having many components involved. To determine role each of these components play, a series of samples—called controls—need to be made and tested along with the sample containing all of components of the remediation technique. Each control sample will test the effectiveness of just one or two of the components needed in the remediation process.

Background Information

Environmental remediation of polluted land, water, and air is a challenge that scientists and community leaders are addressing in order to increase the quality of all life. Whether the pollution is from an oil spill along the Alaskan coast or polychlorinated biphenyls in interior waterways of the United States, cleaning polluted water systems is an expensive and difficult undertaking. Researchers are constantly developing new technologies to make the task of environmental remediation easier and more cost-effective.

One approach to environmental remediation is to introduce a chemical that reacts with the polluting compounds into a contaminated area. The difficulty is twofold; first, selecting a chemical that is not more dangerous than the original pollution, and second, whether the products of the reaction are environmentally friendly. One compound that has shown promise for assisting in environmental remediation is titanium dioxide, TiO_2. Titanium dioxide is a common compound found in white paint and cosmetics. When titanium dioxide is exposed to sunlight, it becomes reactive and capable of oxidizing other compounds.

Within the environment, almost all compounds will eventually degrade. Degradation arises when UV radiation from the sun breaks bonds and causes reactions with chemicals from the surrounding environment. Through these mechanisms, larger more complex compounds are broken down into simpler compounds like metal oxides, carbon dioxide, or water. This process is called **mineralization**. What makes many of the chemicals responsible for pollution so damaging is their resistance to the natural mineralization process.

The role titanium dioxide plays in the environmental remediation of contaminated waters is to accelerate the natural mineralization process. In this way, titanium dioxide is a catalyst. However, titanium dioxide does not react without being activated by sunlight.

[1] Jardim, W.F., Nogueira, R.F.P., *Journal of Chemical Education*, *70*, 1993, p. 861.

Therefore the mineralization of contaminants is **photocatalyzed** by titanium dioxide, meaning that sunlight is a component of the reaction.

In order to determine the rate at which this process takes place, a series of methylene blue standards is used for comparison. Six methylene blue standards with relative concentrations of methylene blue ranging from 100 to 0 will provide references for determining the approximate relative concentration of methylene blue remaining in solution at different times in the mineralization process. Also, in order to determine what components are required in the reaction, several control samples will be created. The control samples all have the contaminant, methylene blue. However, in order to determine how changes in reaction conditions can affect the extent of mineralization, the methylene blue solutions will be treated with either titanium dioxide or sunlight. By separating the reactants into individual tests, a more definitive conclusion can be reached on the effectiveness of this potential environmental remediation technique.

Chemicals and Equipment

- 10 test tubes
- test tube rack
- ~40 mL of 0.1% methylene blue solution
- ~10 mL of 5 g/L titanium dioxide suspension
- 10-mL graduated cylinder
- black paper

Safety Concerns

- **Goggles should be worn at all times.**

Procedure

Preparation of methylene blue standards

Prepare six test tubes using the following amounts of methylene blue solution and deionized water. Label each test tube accordingly for easy reference.

Note: Because each of the experimental and control samples will have 1 mL of either TiO_2 solution or water added, each standard has an extra 1 mL of water.

Test tube #1: 100 methylene blue

Add 5 mL of methylene blue solution and 1 mL of water.

Test tube #2: 80 methylene blue

Add 4 mL of methylene blue solution and 2 mL of water

Test tube #3: 60 methylene blue

Add 3 mL of methylene blue solution and 3 mL of water

Test tube #4: 40 methylene blue

Add 2 mL of methylene blue solution and 4 mL of water

Test tube #5: 20 methylene blue

Add 1 mL of methylene blue solution and 5 mL of water

Test tube #6: 0 methylene blue

Add 0 mL of methylene blue solution and 6 mL of water

Preparation of contaminated samples and controls

Prepare four test tubes using the following amounts of methylene blue solution, TiO_2, and deionized water. Label each test tube accordingly for easy reference.

Test tube #7: Experimental #1

Add 5 mL of methylene blue solution and 1 mL of TiO_2 solution.

Test tube #8: Control #1

Add 5 mL of methylene blue solution and 1 mL of water.

Test tube #9: Control #2

Add 5 mL of methylene blue solution and 1 mL of TiO_2 solution.

Wrap this sample with black paper to isolate it from the sun, but allow it to warm.

Test tube #10: Control #3

Add 5 mL of methylene blue solution and 1 mL of TiO_2 solution.

This solution will be isolated from the sun with the standard solutions.

Initial concentration reading

For each of test tubes #7–10, make a concentration reading by holding the test tube up to the test tube rack holding the methylene blue standards. Approximate the relative concentrations of the methylene blue samples. Record the concentration in the data table.

Reaction time

Place the standards and test tube #10 in the dark to eliminate exposure to the sun. Place test tubes #7, #8, and #9 on a window sill exposed to sunlight. If sunlight is not available, a black light can be used as a substitute.

Take a concentration reading every ten minutes, as you did in the initial concentration reading. Make sure to bring the test tubes from the sunlight into the room to make the comparisons. This will limit the exposure to sunlight for the standards.

After 1 hour, the experiment is concluded and you can dispose of the solutions as directed by your instructor.

Cleanup and Disposal

- ☞ Be sure to wash thoroughly all of the equipment and supplies used during this lab.
- ☞ All solutions should be disposed of in the appropriately labeled waste containers.
- ☞ Be sure to wash your hands thoroughly at the end of this lab.

Data and Observations

Data table

Concentration Readings	Test tube #7	Test tube #8	Test tube #9	Test tube #10
Initial				
10 minutes				
20 minutes				
30 minutes				
40 minutes				
50 minutes				
60 minutes				

Post Experiment Questions

1. What is the purpose of having test tubes #8–10? What information is gained from each?
2. Make a graph of time versus concentration. Plot all of the data for test tubes #7–10 on the same graph.
3. What influence does the sun have on the TiO_2-catalyzed mineralization of methylene blue?
4. What would be the potential benefits and risks if TiO_2 were dumped into a contaminated lake? How could TiO_2 be used in a manner that minimized the risks to the environment?
5. Propose an experiment to study the effects of TiO_2 concentration on the rate of mineralization of methylene blue.